Proceedings of the Lebedev Physics Institute
Academy of Sciences of the USSR
Series Editor: N.G. Basov

The Proceedings of the Lebedev Physics Institute of the Academy of Sciences of the USSR

Series Editor: Academician N.G. Basov

Proceedings of the Lebedev Physics Institute
Academy of Sciences of the USSR
Series Editor: N.G. Basov
Volume 175

LUMINESCENCE CENTERS OF RARE EARTH IONS IN CRYSTAL PHOSPHORS

Edited by M.D. Galanin

Translated by Michael L. Allen

NOVA SCIENCE PUBLISHERS

COMMACK

NOVA SCIENCE PUBLISHERS
283 Commack Road
Suite 300
Commack, New York 11725

This book is being published under exclusive English Language rights granted to Nova Science Publishers, Inc. by the All-Union Copyright Agency of the USSR (VAAP).

Library of Congress Catalog Card Number: 88-1545
ISBN 0-941743-10-1

The original Russian-language version of this book was published by Nauka Publishing House in 1986.

Printed in the United States of America

FOREWORD

Rare-earth-activated luminophors may be divided into two classes — classical crystal phosphors with recombination luminescence mechanisms, and anti-Stokes luminophors with cooperative mechanisms. The results of research on these and other luminophors are represented in this collection. Although the flow of physical processes occuring in them between the act of absorbing an excitation photon and the act of emitting a luminescence photon differ substantially, the kinetics of this luminescence has certain common features. Chief among these is dependence of the rate of the energy conversion process on the pumping intensity, which is reflected in the nonlinearity of the corresponding component in the kinetic equations. Since the kinetics of recombination luminescence have already been researched in some detail in numerous papers, we devote little attention to them here — except to note some of the actual properties of the investigated luminophors which are, in part, linked to the complexity in their system of excited levels. This complexity is the second common feature of luminophors activated by trivalent rare-earth elements, which unifies luminophors with cooperative and recombination luminescence mechanisms. It appears in the kinetics as well as the luminescence spectra of both classes of luminophors.

In contrast to recombination crystal phosphors, the luminescence kinetics of cooperative luminophors have clearly been insufficiently studied up to now. Thus the largest article is devoted to this, filling out the first half of this collection. In it are examined the limits in principle on cooperative luminescence efficiency in connection with the existence of interactions between the forward and back electronic transitions, and the role of the co-activator in luminescence is explained, as well as the effect of the interaction between excited rare-earth ions and base lattice vibrations on the quantum yield.

The remaining three articles are devoted in general to studying luminescence centers in single crystals of zinc sulfide activated by rare-earth elements, and fundamental consideration is given to their properties which are linked to the special properties of rare-earth elements as activators. Thus, the second article in size (and order) is devoted in general to interpretation of the luminescence spectra of ZnS:Tm crystals, which at liquid helium temperatures breaks down into some two hundred lines, as well as interpertation of the ways in which the excitation energy reaches the thulium ions. The third article examines the special properties of the kinetics of luminescence polarization in ZnS:Eu and ZnS:Tm crystals in connection with the fact that, because of the Jahn-Teller effect, the trivalent ions of these activators are displaced relative to the center of the tetrahedron formed by the sulfur ions. Finally, in the last article a method is described for doping activators into crystals under the action of an electric field. This method is especially important in activation with rare-earth elements because, due to their large ionic radius, they don't dope well into zinc sulfide crystals by the diffusion method.

As is apparent from this brief review, the articles placed in this collection cover different aspects of luminescence in rare-earth luminophors. However they possess a certain internal unity. Not only are they united by common co-authors, more than that, since the work was carried out in parallel, the results of each was taken into account, either in explicit or implicit form, during all the others.

M. D. Galanin

CONTENTS

Anti-Stokes Radiation
Conversion in Luminophors with
Rare Earth Ions

A.K. Kazaryan, Yu.P. Timofeyev
M.V. Fok

Abstract: A limiting value of the efficiency of direct anti-Stokes conversion of IR radiation to visible light (up to 10% at pumping densities of 1 W/cm^2) is determined based on an analysis of the balance equations. The high efficiency of anti-Stokes luminescence is due to the narrowness of the energy levels of RE^{3+} ions. The role of the sensitizer is to improve the ratio of the coefficients of summation and cross-relaxation of the pumping energy; the value of this coefficient also determines the color of the anti-Stokes emissions of Er^{3+} in various crystal bases. A comparison of the results of the calculation with experiment shows that the efficiency of current anti-Stokes luminophors is no more than an order of magnitude less than that of the optimized model. At pumping densities of less than 0.1 W/cm^2, luminophors with recombination luminescence mechanisms more efficiently convert IR radiation to visible light.

The direct conversion of IR radiation to visible light is possible in a number of rare-earth-ion doped crystal phosphors, contrary to the empirical Stokes' law. Such a significant increase in the radiation frequency is of interest not only for practical applications, but also from the point of view of the kinetics of the corresponding processes of energy summing in the RE^{3+} ions. Thanks to the numerous studies [1-3] leading up to the papers of N. Bloembergen [4], P. P. Feofilov, V. V. Ovsyankin [5-6] and F. Auzel [7,8], it has been unequivocally established at present that the most efficient mechanism for such processes is a sequential sensitization of the radiating ion by another ion absorbing IR radiation. However, an explanation of the principal channels for transfer, summing and losses of energy in actual systems remains a rather complicated problem. This complexity is caused first and foremost by the extremely branched nature of the energy terms in the ions used (Er^{3+}, Ho^{3+}, Tm^{3+}), as well as the fact that their parameters

1

are highly dependent on the crystal matrix and even the conditions of synthesis. Along with this is the fundamental fact that the probabilities of the forward and reverse optical transitions are inseparably coupled with one another, significantly restricting the range of possible values of the maximum efficiency of such anti-Stokes radiation conversion.

In this context, the question arises: what position do the anti-Stokes luminophors occupy among the other classes of luminophors which, by one means or another, enable us to visualize infrared radiation and, what is especially important in practice, what advances can we expect from further development of anti-Stokes luminophors, i. e. how much less will their efficiency be than that of an idealized model? In some small measure to answer these questions, in this article we have conducted an analysis of simplified models of cooperative anti-Stokes luminophors and compared the results of this analysis with the results of an experiment conducted with both Stokes and anti-Stokes pumping.

1. THE FUNDAMENTAL QUESTIONS AND METHODOLOGY OF RESEARCHING ANTI-STOKES LUMINESCENCE OF RE^{3+} IONS

At the present time, several mechanisms are known for summing the energy of simple pumping of RE^{3+} ions that leads to a direct conversion of IR radiation to visible light (Fig. 1). Therefore one of the central problems of the initial investigation consisted of identification of the anti-Stokes conversion mechanism in actual systems and explanation of how that mechanism could insure a higher efficiency of luminescence. Historically, the first to be examined was the mechanism of sequential ("stepwise") absorption of several IR photons in the same rare-earth ion, which thereupon made the transition to a higher energy state. Such a scheme for an IR photon counter, as proposed by N. Bloembergen [4], was also used to interpret the results of the first experiments [9,10] to observe the luminescence of RE^{3+} ions during pumping in the near IR region.

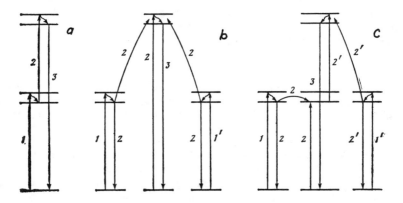

Fig. 1. Mechanism for Summing Ion Pumping Energies   a- Sequential absorption (N. Bloembergen); b- Cooperative sensitization (V. V. Ovsyankin and P. P. Feofilov); c- Sequential sensitization (F. Auzel). The vertical arrows represent optical transitions, the sloping arrows are pumped transfers and the wavy arrows are multiphonon relaxation. The numbers refer to the sequence of transitions; identical numbers refer to all the processes occuring simultaneously.

Table 1. Experimental Criteria for Differentiating the Mechanisms of Anti-Stokes Luminescence in RE^{3+} Ions

Mechanism	Coupling between the absorption $\kappa(\lambda)$ and pumping spectra	Dependence of η^* on ion concentration C for small C	Damping time constant, τ^1
Sequential absorption	$\eta^* \sim \kappa_1(\lambda) \cdot \kappa_2(\lambda)$	$\eta^* \sim C$	τ_2
Cooperative sensitization	$\eta^* \sim \kappa^2(\lambda)$	$\eta^* \sim C^2$	$\tau_s/2$
Sequential sensitization	$\eta^* \sim \kappa^2(\lambda)$	$\eta^* \sim C'^2$	$\bar{\tau}/2$

1 Here τ_2 is the lifetime of the two-photon state, τ_s is the lifetime of the one-photon sensitizer state, and $\bar{\tau}$ is the mean lifetime of the one-photon states of the activator and sensitizer

In the study of this anti-Stokes luminescence, for which the energy deficit between pumping and luminescence photons is much greater than the energy of thermal lattice vibration, a principally new phase was embarked upon beginning in 1966 with independent investigations by Soviet scholars P. P. Feofilov and V. V. Ovsyankin [5,6], and French physicist F. Auzel [7,8]. A substantial increase in the efficiency of anti-Stokes luminescence was obtained in the work of these authors by means of doping with supplementary ion-sensitizers (usually Yb^{3+}). They simultaneously proposed new mechanisms, each with subtle differences of its own, to explain the energy summing in several RE^{3+} ions, namely the sequential ("stepwise") and simultaneous ("cooperative") sensitization of the radiating ion by other IR-radiation absorbing ions. In this process, spectroscopic, energetic, concentration and kinetic criteria were developed for differentiating the aforementioned mechanisms of anti-Stokes pumping. These criteria are displayed in Table 1.

In the table it is shown that the excitation spectrum, i. e. the dependence of the efficiency of anti-Stokes luminescence η^* on the wavelength of pumping light λ may be proportional to either the product of the absorption coefficients $\kappa_1(\lambda)$ and $\kappa_2(\lambda)$ corresponding to two successive transitions, or to the square of the absorption coefficient for transitions from the ground state to a singly-pumped (one-photon) state. For all three mechanisms, the dependence of the luminescence efficiency η^* on the pumping density I_B is linear for small I_B reaching a plateau of constant η^* for high enough values of I_B. Thus this dependence can not serve as a criterion for differentiating the luminescence mechanisms. The kinetic criterion is more well-defined. In accordance with it, after pumping has stopped and after some certain transition period (during which the luminescence intensity may even increase), there follows exponential decay with a time constant having a distinct physical meaning.

The most well-defined criterion is, first and foremost, that of the concentration, which appears during comparison of the processes of energy summing in the same ion with the process of summing during "resonant"

3

energy transfer from other ions (both sensitizers and activators). The distinction between cooperative and sequential sensitization can only be made based on the afterglow (decay) kinetics or on the absolute luminescence efficiency at the given IR pumping density. In this regard, the theoretical work of T. Miyakawa and D. L. Dexter [11,12] becomes important, which developed the Dexter-Förster-Galanin theory of resonant pumping energy transfer [13-16] to cover the case of importance in cooperative luminophors wherein significant mismatch of the interacting RE^{3+} ion transition energies (500 to 3000 cm^{-1}) is compensated by emission or absorption of several phonons. It was learned as a result of their calculations that the efficiency of sequential sensitization can be five orders of magnitude greater than that of cooperative sensitization at low pumping intensities.

Systematic investigations conducted later by many authors [1-3] showed that the sequential sensitization processes observed by F. Auzel occurred in the majority of luminophors with RE^{3+} ions (Er^{3+}, Ho^{3+}, Tm^{3+} and Yb^{3+}). Cooperative sensitization was reliably observed only for the Yb^{3+}-Tb^{3+} ion pair, which lacks a one-photon state of the radiating ion that is close in energy to the excited state of the ion-sensitizer. However, primary credit is given to P. P. Feofilov and V. V. Ovsyankin for discovering a fairly broad class of cooperative phenomena, apart from sequential or cooperative sensitization, including other effects such as the cooperative emission of high energy quanta [1] by two singly-pumped ions (for example, Yb^{3+}). Analogous processes of summing one-photon excitations take place in a number of laser crystals and glasses with RE^{3+} ions [18,19] and, according to the data of several authors [20], in other photosensitive systems as well. All this along with the practical applications of luminophors that convert IR radiation to visible light (visible-band LEDs based on highly efficient GaAs IR diodes and luminescent screens for registering the fields of IR laser radiation) has significantly increased interest in the study of such luminophors. However, until recently there was little concurrence on the fundamental channels for summing or loss of pumping energy in actual systems, and the experimental and theoretical estimates by different authors for the efficiency of such conversions differed by 2 to 3 orders of magnitude.

Kushida, in his theoretical articles [21,22], more correctly studied the term structure of RE^{3+} ions and the processes of back-transfer of energy from the activator to the sensitizer, and thereby obtained significantly less difference between cooperative and sequential sensitization efficiencies than Miyakawa and Dexter in [12]. The results of his calculation of the probability for energy transfer between neighboring RE^{3+} ions (up to 10^8 to 10^9 s^{-1}) were in satisfactory agreement with experiments [23], which were in part dedicated to studying the migration of energy among ions with that same charge, Yb^{3+}. Kushida obtained very good concurrence with several experiments in his estimates of the absolute efficiency of green anti-Stokes emission from Er^{3+} in YE$_3$:Yb, Er^{3+} upon IR pumping with a GaAs diode. However, these estimates were formulated under a number of assumptions, some of which are clearly doubtful.

We should remark that the authors of calculations on the probability of pumping energy transfer and relaxation processes have been practically ignoring an analysis of the kinetic equations of detailed balancing for concentrations of the excited ions. On the other hand, in other papers

4

devoted to an analysis of the kinetic equations [24,25], which were set down in more general form for the case of anti-Stokes luminescence in RE^{3+} ions by Mita [26], usually not enough consideration is alloted to the numerous values of the transfer probabilities. Moreover, many authors [24,25] limit themselves to a strictly qualitative analysis, not even making an attempt at quantitative estimates. One of few exceptions is the work of I. I. Sergeyev [27-32], in which for the first time a numerical solution was performed on a computer of a fairly complex nonlinear balance equation for excited RE^{3+} ions taking into account the phonon spectrum of actual lattices. In this he managed to achieve fairly good agreement with experiment in regards to the functional dependence of anti-Stokes luminescence on temperature, on the concentrations of activators and sensitizers, and on the pumping intensity. However, the absolute values of the luminescence efficiency which he obtained turned out to be almost five orders of magnitude lower than those experimentally obtained in more efficient systems. The fundamental reason for such a discrepancy apparently consists of an incorrect choice of numerical values for the effective IR-absorption cross-section and for the probabilities of radiative and multiphonon transfers in the systems under consideration. At the outset of this study there were also very substantial disagreements between various authors, in their experimental data as well as in the interpretation of them. For example, there have been six different schemes proposed [2] to explain the origin of the red emission of Er^{3+} ions ($^4F_{9/2} \to {}^4I_{15/2}$ transition) when they are excited via Yb^{3+} ions (Fig. 2). We will not give a review here of all these various schemes, which include schemes for summing two- as well as three-photon excitations of Yb^{3+} ions (the $^2F_{5/2} \leftrightarrows {}^2F_{7/2}$ transition). We will adduce a more detailed scheme of the primary channels for summing and relaxation of energy for the other (green) emission of Er^{3+} (Fig. 3), for which the majority of authors are of like mind in regards to its origins. The green band of radiation from Er^{3+} is not single-photon (transitions from $^2H_{9/2}$ and $^4S_{3/2}$ to $^4I_{15/2}$), but in several cases one can observe in almost the same region of the spectrum still another emission band (the $^2H_{9/2} \to {}^4I_{13/2}$ transition), which arises from the summing of a large number of single excitations. The proportion between the red and the aforementioned green emission bands differs greatly from one base to the next, and the reason for this has been interpreted differently by various authors.

Along with the objective difficulties (the branching system of terms for RE^{3+} ions and the significant effect of the crystal lattice on their parameters), an important reason for these discrepancies is apparent from the fact that the researchers usually limited themselves to applying either experimental or theoretical methodology. In this connection, for a more precise qualitative, and thereby quantitative, description of these phenomena, we urgently require the integrated application of several methods. In reality, both spectral (determining the energy differences of transitions occuring in part through lattice phonons) and kinetic measurements that allow us to determine the summed transition probabilities are, in like manner, necessary for a reliable estimate of the probability of intra-center and inter-ion transitions. Only by taking into account this kind of data can we expect agreement between calculated efficiencies for anti-Stokes (as well as Stokes) luminescence and the corresponding experimental data.

Measurements of the spectral-energetic and inertial properties of anti-Stokes luminescence of RE^{3+} ions was conducted chiefly using well-known

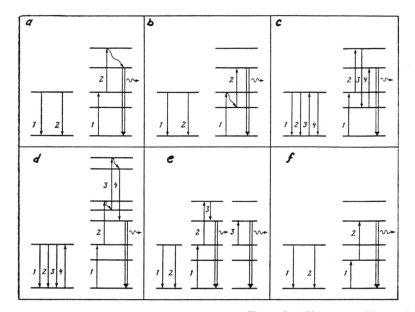

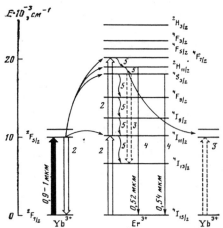

Fig. 2. Various Channels of Anti-Stokes Excitation of the Red Emission Band of Er^{3+} ($^4F_{9/2} \rightarrow {}^4I_{15/2}$ transition) According to [2]. For simplicity of the diagram, absorptions of light by sensitizers are not shown, and the numbers refer only to excitation transfer processes.

Fig. 3. Primary Channels for Cooperative and Cross-Relaxation Energy for the Green Emission Band of Er^{3+} Ions during Anti-Stokes Excitation via Sensitizer Ions Yb^{3+}. 1- Absorption of pumping radiation; 2- Transfer and cooperation of the pumping energy; 3- inter-ion cross-relaxation; 4- Radiative transitions; 5- Multiphonon relaxation

methods using standard equipment. Optical and near-IR pumping of the sample was accomplished with a SIRSh-8-200 spectroscopic incandescent lamp in combination with an MDR-2 high-transmission monochromator; in many of the experiments, including those for measuring afterglow (decay) kinetics, a solid-state (YAG:Nd, $\lambda = 1.06$ and 0.53 μm) or semiconductor (GaAs, $\lambda = 0.91$ μm) laser was used for pumping. Germanium (type FD5G) and silicon (type FD-24K) photodiodes were used as radiation detectors, as well as various photoelectron multipliers, some of which were capable of counting single photoelectrons. Low frequency modulation ($f = 21$ Hz) of the pumping radiation was accomplished in this series of experiments with lock-in amplification (up to 10^5) of the photocurrent and parallel output of data to a DVM and a recorder.

We shall discuss in detail only the methods for absolute measurements

6

of the quantum yield of cooperative luminophors. These measurements were conducted in an integrating sphere with several variants of the different methods first proposed by Z. L. Morgenshtern and V. B. Neustruyev for determining the resonance radiation yield in (R-line) single crystals of ruby [33]. Analogous methods were utilized in a series of articles [34-36] that measured the photoluminescence yield of other low-dispersion media — laser crystals and glasses with RE^{3+} ions or dye solutions. Four independent readings of the luminous flux are required, as registered by a photodetector mounted at the sphere's exit window (Fig. 4), for a determination of the quantum yield η: with the sample near the pumping beam (in position 3) and in it (position 3'), without a diffusing screen (with the screen in position 5) and with the screen located behind the sample (in position 5'). The value of η is computed from this by the following formula:

$$\eta = \frac{l - p - rq}{1 - p - r} \rho_{\mathfrak{z}} \frac{S_{\text{в}}}{S_{\text{л}}} , \qquad (1)$$

where l is the ratio between the photodetector readings with the sample and without the sample, and with the diffusing screen, p is the transmissivity of the sample, i. e. the ratio of readings with and without the sample in the absence of the diffusing screen, r is the specular reflection coefficient of the sample, q is the proportion of reflected light remaining within the sphere, $\rho_{\mathfrak{z}}$ is the diffuse reflection coefficient of the screen, and $S_{\text{в}}/S_{\text{л}}$ is the ratio of the apparatus' sensitivities to pumping (в) and luminescence (л) radiation.

The highest accuracy for this method (within 1 to 2%) is ensured for $S_{\text{в}}/S_{\text{л}} \approx 1$, $\rho_{\mathfrak{z}} \approx 1$, $\eta \approx 1$ and $p + r < 0.5$, i. e. when the quantum losses are small in a sample with significant absorption.

Measurements were conducted in our experiments over a wide range of the spectrum (0.4 to 1.6 μm) on polycrystalline diffuse objects, which required several changes in the measurement set-up. Two variant set-ups of the basic apparatus were employed (Figs. 5,6) which allowed us to determine the quantum yield for resonance radiation η_{p} (i. e., on the wavelength closest to that of the optical pumping) as well as the integrated luminescence $\eta_{\text{и}}$, i. e. the summed quantum yield of all optical and IR band luminescences. The luminescence branching coefficient was measured by an analogous method, i.e. the intensity ratios of individual luminescence bands at a given wavelength of pumping radiation.

For measurements of η_{p} the integrating sphere was set up between two monochromators (Fig. 5) which allowed us to make separate measurements of pumping radiation that had not been completely absorbed by the luminophor screen (due to diffuse reflection and transmission) and the photoluminescence. With a view towards the significant attenuation of the optical flux in such a set-up, we employed an FEU-79 [photoelectron multiplier tube] as the photodetector, which had an amplitude discriminator controlled by high-speed solid-state integrated circuitry allowing measurements of the mean pulse repetition frequency of single photoelectrons (from 10^2 to 10^6 Hz). The need for the first monochromator was dropped during monochromatic laser pumping (in the 0.91, 1.06 and 0.53 μm range), and we managed to conduct the measurements with just a photodiode.

The second monochromator was, obviously, not required for measurements of the integrated quantum yield (except for measurements of sample transmissivity), and so the experiments were usually conducted with

7

Ge photodiodes. These photodiodes had almost constant quantum sensitivity throughout the most important regions of the spectrum in our experiment (0.4 to 1.5 μm) (Fig. 7).

In all these experiments, particular care was taken in the calibration of the spectral sensitivity of the apparatus, including the selective transmission of the photometric sphere. If we consider that, after multiple reflections, the optical flux density over the entire surface of the sphere is practically

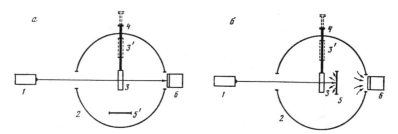

Fig. 4. Set-up for Measuring Photoluminescence Quantum Yield in an Integrating Sphere a- Measuring transmissivity (T); b- Measuring the luminescence quantum yield; 1- pumping source; 2- integrating sphere; 3, 3'- sample; 4- positioning device, inserting and withdrawing the sample from the beam; 5, 5'- diffuse screen; 6- photodetector. Arrows denote the path of the beams.

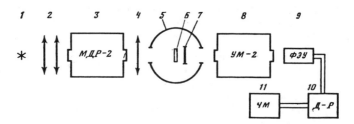

Fig. 5.   Set-up for Measuring Quantum Yield and Photoluminescence Branching Coefficient in Polycrystalline Samples of Cooperative Luminophors 1- Pumping source; 2- condenser; 3- entrance monochromator; 4- focusing lens; 5- integrating sphere; 6- sample; 7- diffuse screen; 8- exit monochromator; 9- photoelectron multiplier tube; 10- photoelectron amplitude discriminator; 11- frequency meter

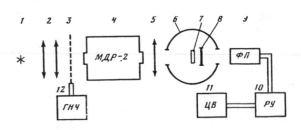

Fig. 6. Set-up for Measuring Integral Photoluminescence Yield

1- pumping source; 2- condenser; 3- modulator; 4- monochromator; 5- focusing lens; 6- integrating sphere; 7- sample; 8- diffuse screen; 9- photodetector; lock-in amplifier; 11- digital voltmeter (or recorder); 12- low frequency generator

8

constant, then it is a simple matter to obtain the following expression for the transmission coefficient Θ (in the presence of the screen):

$$\Theta = \frac{\rho_\vartheta \rho_c}{1 - \rho_c(1 - 2t_0/s)} \frac{s_0}{s}, \qquad (2)$$

where s is the total area of the sphere, s_0 is the area of the entrance and exit windows, and $\rho_c \approx \rho_\vartheta$ are the diffuse reflection coefficients of the walls and screen. It follows from this formula that, for small s_0/s and a sufficiently large value of ρ_c, even small changes in ρ_c (including its variation for different wavelengths) can cause the value of Θ to vary widely. For example, for typical values of $s_0/s = 0.01$ and $\rho_c = \rho_\vartheta = 0.96$, decreasing ρ_c by only 1% causes a 17% drop in Θ, significantly exceeding the limits of experimental error which are no more than 5%.

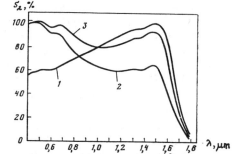

Fig. 7. Spectral Sensitivity of the Apparatus for Measurements of Quantum Yield in a Photodiode-Equipped Integrating Sphere 1- with diffuse screen and uncoated polished walls; 2- coated with a thick layer of MgO; 3- coated with a thin layer of MgO

Inasmuch as the diffuse reflection coefficient is dependent on the thickness of the diffusing layer, and the way it was deposited was changed during the course of its implementation, we systematically conducted measurements of the transmission spectrum of the sphere before each series of experiments. We used a thin coating of magnesium oxide on the specularly reflecting metal surfaces of the sphere to flatten out the spectral sensitivity curve of the apparatus in the visible and near-IR regions. Spectral calibration of the apparatus was accomplished with nonselective thermal radiation detectors (IMO-2 laser calorimeter, RTN-10 thermocouples, etc.). These same detectors were applied in determining the absolute pumping intensity, which is necessary for measurement of the quantum yield of anti-Stokes luminescence η_{ac}, since it depends on the excitation density.

The luminescence yield is computed from a formula similar to (1), except that all measurements are made with the optical diffusing screen set up behind the sample under study. In order to determine the fraction of light absorbed by the luminophor $(1 - p - r)$, the ratio of photodiode readings was taken for the sample in and out of the beam, during which the entrance and exit monochromators were set up to pass only a single wavelength of pumping light. Quantum yield measurements were made with a relatively thin (100 to 200 μm) layer of luminophor, pressed between two glass plates, to insure sufficiently high absorption of the pumping radiation and yet, insofar as possible, to prevent reabsorption of the luminescent radiation. We used different thicknesses for this layer, including one that was practically infinitely thick $(d = 2$ mm), and much thinner luminophor layers deposited on metal substrates or dielectric mirrors. The latter enables a significant (several hundred percent) increase in the luminescence brightness, due to the nonlinear dependence of the intensity of anti-Stokes luminescence on the spatial density of pumping. We used as samples for our research oxychloride,

9

oxysulfide and fluoride luminophors synthesized at FIAN [Lebedev Physics Institute] under the supervision of S. A. Fridman. Several experiments were conducted with the commercially-available anti-Stokes luminophor L-44, as well as other samples. We will not consider the technology of their manufacture in this article, except to note that their base purity with respect to other rare-earth ions or quenching impurities (like Dy or Fe) is usually on the order of 99.9999%.

2. KINETIC LIMITATIONS ON THE EFFICIENCY OF ANTI-STOKES CONVERSION OF IR RADIATION TO VISIBLE LIGHT

In this section, starting from the kinetic balance equations and using the modern theory of transfer, summation and relaxation of electronic excitation of TR^{3+} ions, we will estimate the limiting efficiency of conversion of IR-radiation to visible light. We will investigate the simplest (three- and four-level) model systems and try to make a minimum number of assumptions about the relationships of the probabilities of various processes and the relative population of energy levels in the participating ions. The results of this analysis will be further compared with experimental data for more efficient cooperative luminophors (cf. section 3).

Various authors [21-32, 37-39] have already made theoretical estimates of the efficiency of anti-Stokes conversion. However, the complex nature of the detailed balancing of excited ions required supplementary assumptions, some of which were not always justifiable or correct, or else the use of numerical solutions an a computer. As such, the interaction of various optical characteristics of the participating ions were not taken into account, giving the illusion that a more felicitous choice of these parameters might result in a substantial increase in the efficiency of anti-Stokes luminescence. Meanwhile, by taking the interaction of these parameters into account (for example, the coupling between the effective absorption cross-section and the radiative lifetime for identical transitions), the range of possible values of the efficiency was significantly reduced, compelling us to reexamine the question of the role of sensitizers in increasing the efficiency. We used a method in this calculation similar to that in our earlier papers [40,41] which examined the recombination radiation yield, namely: the unknown quantum yield η is substituted directly into the initial equations for detailed balancing. This allows us to obtain equations in algebraic form for suitably general cases commensurate with the finding of probabilities of various processes.

2.1 Three-level Model of Radiative Centers. We look first into the simplest three-level system that allows conversion of IR-radiation to visible light by summation of the pumping energy of two RE^{3+} ions. This model, first proposed by F. Auzel [7], can be used for a first approximation calculation of the quantum yield η of the fundamental green anti-Stokes emission band of Er^{3+} ions (the $^4S_{3/2} \rightarrow {}^4I_{15/2}$ transition). The only, albeit fairly broad, assumption that we will make consists, first of all, of this, that all activator ions, as well as sensitizer ions, have strictly identical parameters and, secondly, that the ensemble of activator and sensitizer ions complies with the condition that interactions in intermediate excited states (the $^2F_{5/2} \leftrightarrows {}^2F_{7/2}$ transition in Yb^{3+} and the $^4I_{15/2} \leftrightarrows {}^4I_{11/2}$ transition in Er^{3+}) are highly incoherent. These conditions amount to stipulating that a single excitation is distributed with equal probability among all the ions, activators

10

and sensitizers, but that the transfer of energy is accomplished only after relaxation involving the Stark components of the excited state of each ion. Quantitatively this requires that

$$1/\tau_i \ll w_i \ll \Gamma_i \tag{3}$$

where τ_i is the lifetime of such states, w_i is the transition probability, and Γ_i is the transition linewidth (s^{-1}).

Such conditions are met fairly well in efficient anti-Stokes luminophors in that they are doped with rather high concentrations of participating ions (from one to several tens of percentage points) which have, however, forbidden transitions such that the value of w_i is 10^6 to 10^8 s^{-1}, then $10^{-2} \geq \tau_i \geq 10^{-4}$ s and $10^{11} \leq \Gamma_i \leq 10^{13}$ s^{-1} [42-46].

Besides which, it is known from experimental data that increasing the concentration of these ions does not lead to appreciable change in the fine structure of their optical spectra [47], i. e. formation of other types of centers. In this case, and in contradistinction to laser crystals for cooperative luminophors, it is unnecessary, at least to a first approximation, to introduce "microparameters" for the various centers and investigate processes of pumping energy diffusion through lattice sites occupied by participating ions [3,42,46,48]. It is with just this very basic assumption that we may write the kinetic equations of balance for the concentration of excited ions (Fig. 8) in terms of macroscopically-observable averaged parameters. For the investigated systems these equations[1] for steady-state pumping conditions take on the following form:

$$\sigma_S (C_S - N_S) I - N_S/\tau_S - P_{0i} N_S (C_A - N_1 - N_2) + \\ + P_{10} N_1 (C_S - N_S) - P_S N_S N_1 + Q_S N_2 (C_S - N_S) = 0, \tag{4}$$

$$\sigma_A (C_A - N_1 - N_2) I - N_1/\tau_1 - P_{10} N_1 (C_S - N_S) + \\ + P_{0i} N_S (C_A - N_1 - N_2) - 2 P_A N_1^2 + 2 Q_A N_2 (C_A - N_1 - N_2) - \\ - P_S N_S N_1 + Q_S N_2 (C_S - N_S) + \gamma N_2 = 0, \tag{5}$$

$$P_A N_1^2 + P_S N_S N_1 - N_2/\tau_2 - Q_A N_2 (C_A - N_1 - N_2) - Q_S N_2 (C_S - N_S) = 0. \tag{6}$$

Here and henceforth we will use the following conventions: C_S and C_A are the total concentration of activator and sensitizer ions, N_S, N_1 and N_2 are, respectively, the concentrations of sensitizer ions in singly-pumped (one-photon) states, and of activator ions in one- and two-photon states (cm^{-3}); τ_S, τ_1 and τ_2 are the lifetimes of these states resulting from both radiative and radiationless intra-center transitions (s^{-1}), τ_S^*, τ_1^* and τ_2^* are the inverse values of the probability of radiationless transitions from these states, σ_A and σ_S are the effective cross-sections for absorption of pumping IR-radiation (cm^2), I is its intensity (photons·$cm^{-2}s^{-1}$), P_{01} and P_{10} are the energy transfer coefficients from sensitizer to activator ($_{01}$) and the reverse ($_{10}$) in the first excited state, P_A and P_S are the coefficients of summing the energy of two singly-pumped activator ions ($_A$) and the same, with the participation of sensitizer ions ($_S$), and Q_S and Q_A are the coefficients for cross-relaxation of the doubly-pumped (two-photon) activator state producing a one-photon state assisted by a second ion (activator or sensitizer).

---

[1] The factor of 2 in these equations reflects the formation of two single-photon excitations during the decay of one two-photon state.

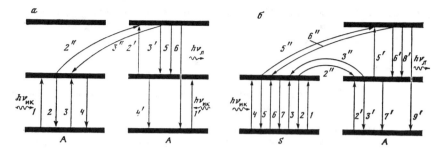

Fig. 8. Diagram of IR-to-Visible Radiation Conversion in a Three-Level Model of Activator Ions (A) Taking Into Account Various Channels for Intra-Center and Inter-Ion Relaxation of Electron Excitation   a- Sysytem without sensitizers (two-stage excitation);   b- system with sensitizers (S) (sequential sensitization).   The numbers denote the sequence od transitions, identical numbers corresponding to processes ocurring simultaneously. The width of the horizontal lines symbolizes the possible splitting of the corresponding lines.

The quantity γ is the probability of intra-center transition of an activator from a state which would cause it to emit a visible light photon, to its one-photon state. These transitions can take place radiatively ($\gamma_{\text{изл}}$), and as the result of multiphonon relaxation (γ_{Φ}) through a system of intermediate levels ($^4F_{9/2}$ and $^4I_{9/2}$ for Er^{3+} ions). In this manner,

$$\frac{1}{\tau_2} = \frac{1}{\tau_2^*} + \gamma_{\text{изл}} + \gamma_{\Phi}, \qquad (7)$$

in which it is necessary that $\gamma_{\text{изл}} \neq 0$ in order to carry out the energy summing, since the probability of energy transfer to a singly-pumped ion is proportional to the probability of this intra-center transition.

In order to gain a more intuitive understanding of the combined parameters that determine the value of η, and more precisely define the role of the sensitizer in increasing it, we would do well to first look at a system without sensitizers, where the IR-radiation is absorbed by the radiating ions themselves (the $^4I_{15/2} \rightarrow {}^4I_{11/2}$ transition in Er^{3+}) and neglect activator-sensitizer interaction processes.

Setting the corresponding terms to zero and combining Eq. (5) with Eq. (6) multiplied by a factor of 2, we obtain the following equation of balance for dissipative transitions in this system:

$$N_1/\tau_1 + 2N_2/\tau_2^* + \gamma N_2 = \sigma_A \left(C_A - N_1 - N_2\right) I. \qquad (8)$$

For a thin luminophor layer, the quantum yield of anti-Stokes luminescence is equal to

$$\eta = \frac{N_2}{\tau_2^*} \frac{1}{\sigma_A \left(C_A - N_1 - N_2\right) I}, \qquad (9)$$

or, factoring in Eq. (7),

$$\eta = \frac{P_A N_1^2}{1 + Q_A \left(C_A - N_1 - N_2\right) \tau_2} \frac{\tau_2}{\tau_2^*} \frac{1}{\sigma_A \left(C_A - N_1 - N_2\right) I}. \qquad (10)$$

12

To discover the relationship of η to the center parameters and the pumping intensity, we must find N_1 and N_2 as functions of I and η. Towards this, we substitute Eq. (9) into the general transition balance equation (8) and obtain the following:

$$\frac{N_1}{C_A\left(1-\frac{N_1+N_2}{C_A}\right)}=\sigma_A\tau_1 I\,[1-\eta(2+\gamma\tau_2^*)]. \tag{11}$$

Since the left-hand side of Eq. (11) is positive, the limiting value η_M for high pumping densities is

$$\eta_M \leqslant \frac{2}{2+\gamma\tau_2^*}\,. \tag{12}$$

This approaches $1/2$ only if $\gamma\tau_2^* \ll 1$. If, as usually is the case in actual luminophors, $\gamma\tau_2^* \gtrsim 1$, then $\eta_M \leq 1/3$, which is in reasonable agreement with the limiting efficiency of anti-Stokes luminophors obtained during laser pumping ($\eta \leq 24\%$) [49].

Similarly, from Eq. (9) we get

$$\frac{N_2}{C_A\left(1-\frac{N_1+N_2}{C_A}\right)}=\sigma_A\tau_2^* I\eta. \tag{13}$$

From Eqs. (11) and (13) it follows that the concentration ratio of activator ions in states of double and single excitation energies is determined by the anti-Stokes luminescence yield and the lifetimes of these states. In this manner, for high values we can expect inverse population of levels during anti-Stokes pumping, which agrees qualitatively with the observed laser effect in anti-Stokes crystals during IR pumping [18,19]. The ratio N_2/N_1 thus obtained allows us to express N_2 and N_1 in terms of paramters of the system η and I, which, on the basis of Eq. (9), gives the following, linking η and I:

$$\frac{\eta}{[1-\eta(2+\gamma\tau_2^*)]^2}=\frac{P_A\tau_1 N_0\,(\tau_2/\tau_2^*)}{1+Q_A N_0\tau_2}\,\sigma_A\tau_1 I, \tag{14}$$

where $N_0=C_A-N_1-N_2$ is the concentration of ions in the ground state with allowance for its depopulation by IR-radiation. This concentration can be determined by adding expressions (11) and (12) and taking into account the fact that their denominators contain the term $(N_1+N_2)/C_A$. In this way, after some simple manipulations we obtain

$$N_0=C_A\left(1-\frac{N_1+N_2}{C_A}\right)=\frac{C_A}{1+\sigma_A\tau_1 I\,[1-\eta(2-\gamma\tau_2^*-\tau_2^*/\tau_1)]}\,. \tag{15}$$

In the most general case, for which the limiting value of η_M is achieved at roughly the same range of intensities I for which there is already significant phototropism of the luminophor relative to IR-radiation, both these effects act on each other in a nonlinear way. This leads to a rather unwieldy expression for $\eta\varphi(\eta)=DI/I_0$, to wit:

$$\eta/\{[1-(2+\gamma\tau_2^*)\,\eta]^2-\eta\left[1-\left(2+\gamma\tau_2^*-\frac{\tau_2^*}{\tau_1}\right)\eta\right]\frac{\tau_2^*}{\tau_2}\,(P_A\tau_1 C_A)^{-1}\}=$$

$$= \frac{P_A \tau_1 C_A (\tau_2/\tau_2^*)}{1 + Q_A \tau_2 C_A} \sigma_A \tau_1 I = DI/I_0, \tag{16}$$

where $I_0 = (\sigma_A \tau_1)^{-1}$. Although this equation is only quadratic in η, it is still tedious to analyze. It is much simpler in using this equation to calculate the pumping intensity required for a given η value, then draw a graph or make up a table of $\eta(I)$ for various luminophor parameters, assuming that $\eta \ll 1$. Such a dependence of η on the normalized pumping intensity I/I_0 is presented

Fig. 9. Dependence of the Quantum Yield of Anti-Stokes Luminescence η on Normalized Intensity of IR-Pumping I/I_0 ($I_0 = 5 \cdot 10^{22}$ s$^{-1} \cdot$cm^{-2}) for Various Ratios of Ion Parameters

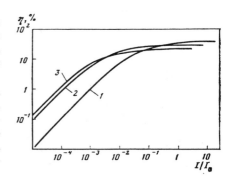

1- $P/Q = 1$, $\tau_\alpha/\tau_2 = 10$, $\gamma\tau_2 = 0.2$; 2- $P/Q = 10$, $\tau_\alpha/\tau_2 = 10$, $\gamma\tau_2 = 1$; 3- $P/Q = 10$, $\tau_\alpha/\tau_2 = 20$, $\gamma\tau_2 = 2$, where τ_α and τ_2 are the lifetimes of the one- and two-photon states relative to optical transitions (s), γ is the probability of intra-center multiphonon relaxation (s^{-1}), and P and Q are coefficients of cooperation and cross-relaxation

in Fig. 9 for various combinations of center parameters. It is apparent from the figure that the transition from a linear dependence $\eta(I/I_0)$ to the limiting constant value η_M is exceedingly gradual, and as such, luminophors of lower luminescence efficiency for small I can have higher values of η for large I.

It is obvious from Eq. (15) that the quantity I_0 represents that pumping intensity necessary for noticeable depletion of the ground state if η is small. Therefore, as follows from Eq. (14), the question of whether the IR-radiation attenuating affect precedes the plateauing of the anti-Stokes luminescence yield or vice versa can be completely answered by a second composite quantity

$$D_A = \frac{P_A \tau_1 C_A (\tau_2/\tau_2^*)}{1 + Q_A \tau_2 C_A}.$$

It is desirable to increase the composite parameter D_A, from the point of view of increasing luminescence yield at small I and from the point of view of efficiency of η in the plateau region (which should fall off for $\frac{N_1 + N_2}{C_A - N_1 - N_2} \approx 1$ for "thick" layers the same as for "thin" ones because of the reabsorption of anti-Stokes luminescence). For the given ion characteristics, this can be achieved by increasing their concentration, for only by starting with some certain concentration will the cooperation of pumping energy predominate over the subsequent absorption of two IR-photons in isolated ions. Actually, an examination of Bloembergen's scheme, which is similar to that which we have adduced, shows that the luminescence intensity ratio for both processes is simply equal to

$$\frac{\sigma_2}{\sigma_A} \frac{1}{P_A \tau_1 C_A}, \tag{17}$$

where σ_2 is the effective absorption cross-section from the one-photon state. When $\sigma_2 \approx \sigma_A$, the condition $\dfrac{\eta_{\text{cooperative}}}{\eta_{\text{subseq. absorp.}}} > 1$ dictates that $P_A \tau_1 C_A > 1$. This inequality can be satisfied in many cooperative luminophors with even very modest ion parameters: $P_A \approx 10^{-16} \text{cm}^3 \text{s}^{-1}$, $\tau_1 \approx 10^{-3}$ s for $C_A \gtrsim 10^{20} \text{cm}^{-3}$, i. e. at even less than one-percent ion substitution in the lattice. However, the excited state cross-relaxation probability (the term $Q_A C_A \tau_2$ in the denominator of D_A) inevitably grows at the same time as the cooperative processes probability (the numerator of D_A). Therefore at high concentrations C_A we obtain

$$\eta \varphi(\eta) = \frac{P_A}{Q_A} \frac{\tau_1}{\tau_2^*} \frac{I}{I_0} \tag{18}$$

which for the given case, in conformance with Eq. (14), $\eta \varphi(\eta)$ (but not the efficiency!) does not in general depend on the depopulating of the ground state, and the value of τ_1/τ_2^* for actual luminophors without sensitizers is ~ 10.

In like manner, the value $\dfrac{P_A}{Q_A} \dfrac{\tau_1}{\tau_2^*}$ directly determines the fraction of singly-excited ions $\dfrac{N_1}{C_A}$ needed to reach the luminescence yield plateau, for large concentrations (i. e. for $P_A C_A \tau_2 > 1$). Meanwhile, and in conformity with the adduced formulas and Fig. 9, the transition from a quadratic to a linear dependence of the brightness of anti-Stokes luminescence on I must be made very gradually, in contradistinction to the approximations introduced in several experimental papers. For example, when $\gamma \tau_2^* \approx 1$ and $\eta \approx 0.1$, the calculation of I from the approximate formula $\eta = DI/I_0$ (corresponding to a quadratic dependence on brightness) gives a value of I that is roughly a factor of 2 too low, and at $\eta \approx 0.25$ the error reaches almost an order of magnitude.

In other words, increasing η by a factor of 2.5 requires an increase in I by a factor of, not just 2.5, but 12.5, in spite of the fact that η itself is ≤ 0.25.

On top of this, scatter of the center parameters, inhomogeneous pumping depth-wise in the luminophor layer and other secondary effects can only smooth out this transition all the more.

The following rather simple formula is obtained for the luminescence yield at moderate intensities of Stokes pumping (for which cooperation can be neglected),

$$\eta_{\text{CT}} = \frac{\tau_2}{\tau_2^*} \frac{1}{1 + Q_A C_A \tau_2} \cdot \tag{19}$$

15

i. e., the luminescence efficiencies for Stokes and anti-Stokes pumping at moderate intensities[2] are rather simply coupled:

$$\eta/\eta_{CT} = P_A \tau_1 C_A l/I_0 \tag{20}$$

which has received little attention in studies of cooperative luminophors. We will look at the interaction and permissible numerical values for center parameters in a later section, but for now we present the results of a similar examination of sensitizer-ion doped luminophors. Since even in the absence of sensitizers it is possible to get $D \gg 1$, then the η plateau will be reached before $(N_1+N_2)/C_A$ ever gets to $\approx \frac{1}{2}$, and so we will assume $N_0 \approx C_A$. The other important condition, analogous to that of highly incoherent interactions in the activator and sensitizer ion ensembles, is to specify that these very same interactions take place among these ensembles.

Under these conditions the concentration ratio of singly-pumped activator and sensitizer ions is given by the simple formula

$$\frac{N_1}{N_S} = \frac{P_{01}}{P_{10}} \frac{C_A}{C_S} . \tag{21}$$

The luminescence quantum yield of this system is determined to be

$$\eta = \frac{N_2}{\tau_2^* (\sigma_A C_A + \sigma_S C_S) I} = \frac{P_S N_S N_1 + P_A N_1^2}{1 + (Q_S C_S + Q_A C_A) \tau_2} \frac{\tau_2}{\tau_2^*}, \tag{22}$$

and the general equation of balance for the dissipative transitions has the form

$$\frac{N_1}{\tau_1} + \gamma N_2 + \frac{N_S}{\tau_S} + 2\frac{N_2}{\tau_2^*} = (\sigma_S C_S + \sigma_A C_A) I. \tag{23}$$

It is easy to find N_S by substituting the ratio $\frac{N_1}{N_S}$ into this equation:

$$N_S = \frac{[1 - (2 + \gamma \tau_2^*) \eta] (\sigma_S C_S + \sigma_A C_A) \tau_1 I}{1 + \dfrac{P_{01}}{P_{10}} \dfrac{C_A}{C_S} \dfrac{\tau_S}{\tau_1}} . \tag{24}$$

In this manner we obtain a general expression analogous to that given above for $\eta \varphi(\eta) = D_S I/I_0$ in the form

$$\frac{\eta}{[1 - (2 + \gamma \tau_2^*) \eta]^2} =$$
$$= \frac{\left[P_S \dfrac{P_{01}}{P_{10}} \dfrac{C_A}{C_S} + P_A \left(\dfrac{P_{01}}{P_{10}} \dfrac{C_A}{C_S} \right)^2 \right] \tau_S^2 (\sigma C + \sigma_1 C_1) \dfrac{\tau_2}{\tau_2^*} I}{\left(1 + \dfrac{P_{01}}{P_{10}} \dfrac{C_A}{C_S} \dfrac{\tau_S}{\tau_1} \right)^2 [1 + (Q_S C_S + Q_A C_A) \tau_2]} , \tag{25}$$

which may be compared to expression (14) for a luminophor with no sensitizers. Along with the supplementary contribution of the sensitizers in absorption of pumping radiation and cooperation of one- and two-photon excitations, we notice also undesirable effects — an additional decay of the two-photon state and a decreased lifetime of the one-photon state.

---

[2] At high intensities of Stokes pumping, η_{CT} in such systems increases due to cooperation processes

16

As was shown by our experiments in anti-Stokes luminophor afterglow radiation kinetics in the presence of activators and sensitizers, IR luminescence from both types of ions decays with the same time constant [45]

$$\frac{1}{\bar{\tau}} = \frac{\tau_S^{-1} C_S P_{10} + \tau_1^{-1} C_A P_{01}}{C_S P_{10} + C_A P_{01}}, \tag{26}$$

which for optimal concentrations $C_{Yb} \approx 20\%$ and $C_{Er} \approx 3\%$ is very close to the lifetime of a sensitizer ion, and much shorter compared to the activator lifetime ($\bar{\tau} = 0.42$ ms when $\tau_S = 0.40$ ms and $\tau_1 = 2.1$ ms for YOCl). We note that identical decay rates for both ions serves as a clear indicator that conditions of highly incoherent interactions in such systems has been met.

It will be shown in the following section that an increase in the effective cross-section for absorption of pumping radiation is usually fully compensated by a decrease in τ_S relative to τ_1. Therefore the sensitization is stipulated primarily by the ratio of P_S/Q_S relative to P_A/Q_A, which allows us to use higher concentrations of participating ions, thereby also increasing the IR absorption coefficient. We will note also the other merits of sensitized luminophors that appertain to practical applications while still maintaining the aforementioned luminescence yield. Firstly, there is an improvement in the tolerances for this system, both spatial (growth of $\sigma_S C_S$) and temporal (decrease in $\bar{\tau}$). Secondly, there is a decrease in visible light reabsorption which, when the internal quantum yield is different from 1, should lead to a decrease in the external technical luminescence yield (especially for $\tau_2^* \ll \tau_1$).

All the other qualitative conclusions made about luminophors without sensitizers are also valid for sensitized luminophors. If we assume further that $\sigma_A C_A \ll \sigma_S C_S$, and $P_A \ll P_S$ (or $Q_A \gg Q_S$), then for a given sensitizer concentration there must be an optimum activator concentration. In the general case this will depend upon both the variation in energy migration to the activator, and cross-relaxation decay of two-photon states with participation of a second activator ion. When the first effect predominates,

$\left[\dfrac{C_A}{C_S}\right]_{OPT} \approx \dfrac{\tau_1}{\tau_S}\dfrac{P_{10}}{P_{01}}$, i. e. only enough pumping should be directed toward an

activator so that both channels yield the same number of transitions (or the same IR-luminescence yield, if we take into account intra-center losses). If two-photon state decays play the dominant role, then the optimum

situation is determined from the condition that $\left[\dfrac{C_A}{C_S}\right]_{OPT} = \left[2\dfrac{Q_S}{Q_A}\right]^{\frac{1}{2}}$. Inasmuch

as τ_S is significantly less than τ_1, and usually $C_A/C_S < 1$ in experiments, then we should assume that the second case obtains in which, corresponding to our qualitative conclusions, $Q_S/Q_A \ll 1$. These qualitative conclusions were verified with numerical calculations directly from formula (25) computed with various luminescence and sensitization center parameters for three different activator concentrations (Fig. 10). It was in fact derived from these calculations that unless a sensitizer has a better ratio of forward to back transitions than an activator, it will provide practically no gain in the anti-Stokes luminescence yield. On the other hand, by selection of some entirely possible numerical values for the ratio of the cooperation and relaxation coefficients in sensitizers and activators and the ratio of coefficients for one-photon transfer between these ions, we managed to achieve satisfactory

17

agreement with the experimental data (Fig. 10 d). We obtained values that were , for the most part, close to experiment for $(C_S)_{OPT}=19\%$ (at $C_A=3\%$), i.e. $(C_A/C_S)_{OPT}=0.16$ and for the gain in efficiency of η due to the introduction of sensitizers (10^2). In addition, the maximal value of η itself agreed with experiment, being about 2.5% for pumping density $I=5 \cdot 10^{18} cm^{-2}s^{-1}$ (1 W/cm^2). We will turn again to a discussion of these questions when we examine the numerical values of parameters for real systems.

Unfortunately, the formulas we obtained have no optimum, just a plateauing of η, for the concentration function $\eta(C_A, C_S)$ during simultaneous increase of C_A and C_S (except for the obvious one from the condition $C_A + C_S = C_0$, where C_0 is the total number of cations in the lattice). Thus for agreement with the experimental data we need to study either the variation in center parameters at very high temperatures, or the existence of decay processes for two-photon states with the participation of three or more ions (the Dexter process, inverse to the cooperative sensitization process proposed by Feofilov and Ovsyankin to explain anti-Stokes luminescence). Finally, in real luminophors there are uncontrolled quenching impurities (at the very least in concentrations of $10^{-6}C_0$). Quenching takes place when the mean distance between quenching ions is on the order of the

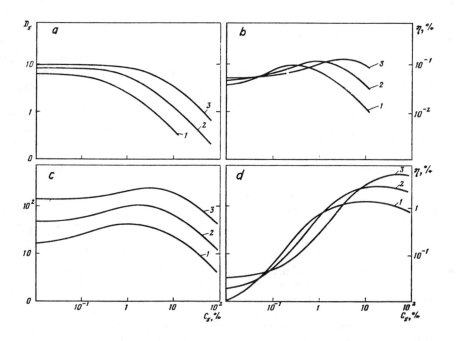

Fig. 10   Dependence of the Composite Parameter D_S and the Anti-Stokes Luminescence Yield η (at IR Pumping Intensity of 1 W/cm^2) on Sensitizer Concentration C_S for Various Ratios Between the Coefficients of Cooperation P and Cross-Relaxation Q, and Between One-photon States P_{SA}/P_{AS} for Three Activator Concentrations C_A (curves 1, 2 and 3 correspond to $C_A=1$, 3 and 10%)

18

radius of the energy migration sphere R_M. Actually, our experiments showed [50] that intentional doping with Dy^{3+} ions ($10^{-4}C_o$) not only strongly quenched the anti-Stokes luminescence, but the maximum of η was also shifted towards the region of lower C (from 22 to 16%).

The radius of the pumping energy migration sphere can be evaluated approximately by the following elementary formula

$$R_{\text{M}} = \left(\frac{v^2 \tau_{\text{п}} \tau}{3} \right)^{1/2} = \left(\frac{R^2 \tau}{3\tau_{\text{п}}} \right)^{1/2}, \tag{27}$$

where R is the distance between adjacent lattice sites (≈ 3.5 Å), $\tau_{\text{п}} \approx 10^{-8}$s is the time for transfer between them, and $\tau \approx 10^{-3}$s is the excited state lifetime. By substituting these values for v, $\tau_{\text{п}}$ and τ we obtain a radius of the migration sphere of about $6 \cdot 10^{-6}$cm. Consequently, the purity of the anti-Stokes luminophors must be on the order of 99.9999%, which was achieved in the manufacture of the luminophors used in our study. The existence of quenchers may be formally introduced into our system of equations as a reduction in the excited state lifetimes, a situation which is in fact observed in the experiment after doping with Dy^{3+} ions.

Several experiments testify to the finiteness of the distance R_M, in which a more-than-quadratic dependence of anti-Stokes luminescence intensity on I was obtained for low pumping intensities. Such a dependence would be observed if the mean distance between singly-pumped ions were greater than R_M. In addition, a highly linear dependendence of η on I over a fairly broad range of pumping intensities and the occurence of quenching for even very small concentrations of quenchers testifies to the fact that R_M is rather large (about 100 lattice constants). This serves as additional justification for the validity of the assumption of highly incoherent interactions between participating ions , and it was within this limitations that we were able to set down and solve the kinetic equations of balnce for the excited particles.

2.2 **Coupling of the Parameters Determining the Efficiency of RE^{3+} Ion Anti-Stokes Luminescence.** The range of variation in the first parameter ($I_0 = (\sigma_A \tau_1)^{-1}$) is set by the fundamental relationship between the Einstein coefficients determining the optical transition probability [51]. In actuality, for the simplest two-level system the product of the maximum effective absorption cross-section σ_M and the spontaneous emission lifetime τ^* are determined by just the wavelength of the emission λ and the spectral width of the transition $\Delta v_{\text{эф}}$, not its probability:

$$\sigma_M \tau^* = \frac{\lambda^2}{8\pi n^2 \Delta v_{\text{эф}}}, \tag{28}$$

where n is the index of refraction, $\Delta v_{\text{эф}}$ is found from the condition $\int \sigma(v)dv = \sigma_M \Delta v_{\text{эф}}$ and is given in units of s^{-1} (obviously, $\Delta v_{\text{эф}} = c\Delta w_{\text{эф}}$ where c is the speed of light and $\Delta w_{\text{эф}}$ is the width of the transition in cm^{-1}).

What is most remarkable is that this quantity does not depend on the resolution or order of the transition or the form of the line, and for a given λ and n only the spectral width of a band is determined. Therefore the very narrowness of RE^{3+} ion absorption bands that makes them preferred accumulators of electronic excitation energy at the same time fundamentally limits the width of the IR pumping band ("resonant systems"). The fact that the value of $\sigma_M\tau^*$ is a constant, which is well-known in quantum electronics [51,52] but not heretofore applied in estimating the efficiency of anti-Stokes luminescence, does not allow us to account for the role of the sensitizer by simply increasing its absorption, because the lifetime of the one-photon state decreases at the very same rate.

Substituting into Eq. (28) the values $\lambda = 0.97$ μm, $n = 1.7$ (from 1.5 to 2.1) for the anti-Stokes luminophor matrix, and a quite narrow effective IR pumping spectrum of $\Delta v_{\text{эфф}} = 200 \text{cm}^{-1}$ (for $T \approx 300$ K), we obtain $\sigma_M\tau^* \approx 2 \cdot 10^{-23} \text{cm}^2 \cdot \text{s}$ or $I_0 = 5 \cdot 10^{22}$ photons$\cdot$cm$^{-2} \cdot$s^{-1}. It follows immediately that if the value of D were on the order of unity or less, then the anti-Stokes luminescence efficiency η^* would not exceed $1 \cdot 10^{-4}$ at the IR pumping densities of $I = 1$ W/cm^2 (or $5 \cdot 10^{18}$ photons$\cdot$cm$^{-2} \cdot$s^{-1}) used in the series of experiments to determine that efficiency. Meanwhile, as follows from our experiments [50] and those of others [2], at such pump powers η is roughly two orders of magnitude higher and so, as we postulated, $D \gg 1$. This conclusion differs sharply from the results of calculations in articles [27-32], and one of the principal reasons for this discrepancy is their use of a value for the effective absorption cross-section of Yb^{3+}, $\sigma_{Yb} = 10^{-21}$cm^2, which is clearly too low when at the same time they make the assumption that the emission time of all transitions is $\tau^* \approx 10^{-3}$ s. On the contrary, with a known value of τ_{Yb}, e. g. $4 \cdot 10^{-4}$s for YOCl, this relation gives $\sigma_M \approx 5 \cdot 10^{-20}$cm^2. This value is in satisfactory agreement with the spatial resolution A_p for thick layers of those luminophors ($A_p \approx 10$ lines/mm) or for optimal thickness of the phototropic layers of the luminophor equal to 0.1 to 0.2 mm [8]. These estimates are particularly critical, insofar as it is difficult to conduct direct measurements of σ in powdered anti-Stokes luminophors.

The adduced estimate for $\sigma_M\tau^*$ on average is in fair agreement also with the results of separate measurements of σ_M and τ^*, including measurements made on laser crystals and glasses activated with Nd^{3+}, Yb^{3+}, Er^{3+} and other RE^{3+} ions. However, there is about an order of magnitude difference between the above value for the product and the maximum and minimum values of the product calculated from experimental data, which was systematized in monographs [51,52]. This difference, in addition to changes in the width of the transition (up to 10 to 20 cm^{-1} at low temperatures) and possible experimental error, is attributable to the following causes: first of all, the experimental values for τ can be both higher (during reabsorption and secondary luminescence in samples with high IR luminescence yields) and lower (from internal and external quenching, as well as radiative transitions to other levels) than the radiative lifetime. Secondly, in highly symmetric matrices and those with small (compared to kT) Stark splitting, one needs to look at the degree of degeneracy in the ground and excited states which, for anti-Stokes luminophors, introduces an additional multiplying factor of somewhere between 0.75 and 0.90 (the $^2F_{5/2}$ and $^2F_{7/2}$, $^4I_{15/2}$,and $^4I_{3/2}$ or $^4I_{11/2}$ states). Finally, in bases with lower symmetry and Stark splitting which is appreciable compared to $kT = 210$ cm^{-1} (for $T = 300$ K), which pertains principally to anti-Stokes luminophors, one should look into the standard deviation of the optical transition probability and thermal population of the ground and excited state Stark components.

It is possible to do all this by analysis of the experimental emission and absorption spectra which is used in part for estimating the luminescence yield [46,51] from the data of spectral and kinetic measurements. The value of η_{CT} obtained in this manner for laser crystals and glasses doped with Nd^{3+} and Er^{3+} agrees to a sufficient accuracy (within 10%) with the results of direct measurement in the integrating sphere [34,35]. However, in anti-Stokes luminophors we are interested primarily in the value of $\sigma_{ij\max}$ (i and j are the numbers of the Stark components of the upper and lower term), which we can calculate in an analogous way if we know the Stokes IR luminescence efficiency $\eta_{ИК}$ and the branching coefficient for luminescence transitions β_{ij}. It is easy to show in this case that

$$\sigma_{ij\max}\tau^* = \frac{\lambda^2}{8\pi n^2 \Delta\nu_{эф}}\,\eta_{ИК}\frac{N_i}{N_j}\,\beta_{ji}, \tag{29}$$

where $\eta_{ИК}$ is the Stokes IR luminescence yield, the branching coefficient β_{ij} is the relative proportion of the given transition in the overall luminescence spectrum,

$$N_i = \left[\exp\left(-\frac{\Delta E_i}{kT}\right)\right]\Big/\sum_k \exp\left(-\frac{\Delta E_k}{kT}\right) \text{ and } N_j = \left[\exp\left(-\frac{\Delta E_j}{kT}\right)\right]\Big/\sum_m \exp\left(-\frac{\Delta E_m}{kT}\right)$$

are the term populations of the Stark components with the provision that $\Sigma N_i = \Sigma N_j = 1$. Based on our experiments (cf. section 4.3) for YOCl:Yb,Er luminophors, $\eta_{ИК} = 0.8$ and $\beta_{ij} \approx 0.6$, and so $\sigma_M \tau \approx 1 \cdot 10^{-23}$ cm^{2}s^{-1}, i. e. roughly a factor of 2 lower than our simplified estimate.

During pumping by non-monochromatic radiation the value of $\bar{\sigma}\tau$ may of course be substantially less (and I_0 be greater). If, for example, the pumping radiation and absorption bands have a Gaussian distribution (over the scale of frequencies) and their maxima coincide, then the corresponding calculation is easily carried out. It gives us a value of

$$\bar{\sigma} = \frac{\sigma_{\max}\Delta\nu_B}{\sqrt{\Delta\nu_{эф}^2 + \Delta\nu_B^2}}, \tag{30}$$

where $\Delta\nu_B$ is the width of the pumping band. In the particular case of $\Delta\nu_B \approx \Delta\nu_{эф}$, the value of $\bar{\sigma}$ is $\sqrt{2}$ times less, i. e. I_0 grows by $\sqrt{2}$, but it becomes lower by a factor governed by the maximum pumping radiation spectral density $I_{\lambda_{\max}}$. Further increase in the width of the pumping band, e.g. by an incandescent lamp, are wholly undesirable, since I_0 grows sharply without a significant gain in $I_{\lambda_{\max}}$. For pumping at the edge of the absorption band (such as from a YAG:Nd laser at $\lambda = 1.06$ μm) at $T = 293$ K, the value of $\sigma\tau$ can be around 30 times lower (due to the weak populating of the upper Stark component of $^2F_{7/2}$) and $\sigma \approx 10^{-21}$cm^2. However, neither the transition fine structure discussed above, nor spectral matching of the pumping will affect the relationship of the absorption saturation effect and the plateauing of η for anti-Stokes transitions, for these are completely determined by the multiplying factor D. This factor alone determines the value of N_1/C_A at which the radiation yield begins to plateau or, more precisely, determines the value of $\dfrac{\eta}{(1-2\eta)^2}$ for a given N_1/C_A.

We turn now to the question of the coefficients of forward and back energy transfer and their connection with the probability of intra-center optical transitions. It follows from the Förster-Dexter-Galanin theory [13-15] that the maximum probability of energy transfer, in the optimal case of no Stokes shift and complete coincidence of the donor emission spectrum and the acceptor absorption spectrum, for the dipole-dipole interaction, is given by

$$w = \frac{3}{4}\left(\frac{\lambda}{2\pi R}\right)^6 \frac{\Delta\tau}{\tau_a \tau_{\text{д}}} = \frac{3}{4}\left(\frac{\lambda}{2\pi R}\right)^6 \frac{1}{2\pi\Delta\nu C}\frac{1}{\tau_a \tau_{\text{д}}}, \tag{31}$$

where R is the distance between ions, τ_a and $\tau_{\text{д}}$ are the excited state lifetimes of the acceptor and donor states, $\Delta\tau$ is the time for relaxation within the width of a level, corresponding to $\dfrac{1}{2\pi\Delta\nu c}$.

Substituting in the values of $\lambda = 1$ μm, $R = 3.5$ Å (the distance between lattice sites), $\Delta\nu = 200$ cm^{-1}, $\tau_{\text{д}} = \tau_a = 10^{-3}$s which are typical of RE^{3+}-ion doped anti-Stokes luminophors, we obtain $w = 1.6\cdot10^8$s^{-1}. This value, obtained for a very simple case, coincides surprisingly well with T. Kushida's estimate $(w = 1.4\cdot10^8$s$^{-1})$ made for the quadrupole-quadrupole interaction with Stark splitting of the RE^{3+} ion levels, and also with experiments [21,22] on energy migration between ions for complete substitution of the lattice cations (10^8 to 10^9s^{-1}).

In light of the transfer probability's very strong dependence on the distance between ions (R^{-l}, where $l = 6$ for the dipole-dipole interaction, 8 for the dipole-quadrupole and 10 for the quadrupole-quadrupole interaction), at high enough concentrations of the participating ions, energy transfer primarily takes place between nearest-neighbor lattice sites. Such conditions are realized in cooperative luminophors, so that the role of the media's polarizability in this case is reflected only indirectly, through the values of τ_a and $\tau_{\text{д}}$, and the index of refraction should not appear all by itself (in the form n^6) in the denominator of the expression for w. An increase in the concentration C of participating ions to a first approximation increases only the number of interacting ions, not the distance between them, which is justification for introducing certain averaged values of the transfer coefficients $\overline{P}_{\text{M}} = w/C$. In this way, and using the above-mentioned value for w, we obtain a rough estimate of the maximum value of $P = 5\cdot10^{-15}$cm$^3\cdot$s^{-1}.

The above estimate is apparently used equivalently for processes of energy transfer from excited to non-excited ions, as well as between excited ions with a transition of one of them from a two-photon state (cooperation), or for reverse processes — inter-ion cross-relaxation of this higher-energy state into two excited states of lower energy. The specific property of rare-earth ions which allows them to store up and cooperate excited states is the brief lifetime of those states, as was considered earlier (the forbiddenness of transitions) [54], and the relatively small level width. In fact the values of $(\sigma_A\tau_1)$, $(P_A\tau_1)$ or $(\sigma_S\tau_S)$, $(P_S\tau_S)$ entering into the numerator of the expression for radiation yield at small l are inversely proportional to the square of the energy level width and the radiative lifetime of the transition of a radiating ion (activator) from a one-photon state to one of lower energy. The dependence of these expressions on the transition wavelength will be mostly the following: λ^6 for dipole-dipole transitions, λ^8 for quadrupole-dipole, and λ^{10} for quadrupole-quadrupole. This is why it is significantly simpler to carry out cooperation of the pumping energy for near-IR pumping than it is to convert visible light to ultraviolet radiation (changing λ by a factor of 2 changes the maximum value of D/I_0 by a factor of 60 to 240!).Therefore the energy summing phenomenon in RE^{3+} ions is in principle more suited for converting IR radiation to visible light or even longer wavelength IR radiation $(\lambda_B = 2$ to 3 μm$)^3$ to IR of a shorter wavelength (e. g. $\lambda\approx1$ μm), within the

---

[3]The long wavelength limit is dictated by multiphonon relaxation.

sensitivity limits of ordinary photodetectors. At present there are no cooperative luminophors with sensitivities beyond $\lambda \geq 2$ μm, but anti-Stokes conversion is noticeably facilitated for $\lambda_B = 1.5$ to 1.6 μm, so that fairly efficient summing of two, three, and even four excitations can take place.

Considering the values derived above for $w \approx 1 \cdot 10^8$ s^{-1}, $C_A/C_0 \approx 0.2$, $\tau \approx 10^{-3}$ s, $\sigma \tau \approx 10^{-23} cm^2 \cdot s^{-1}$ (for $\lambda = 1$ μm), we obtain the result that the maximum value of the numerator of D can be $2 \cdot 10^4 \gg 1$ and correspondingly that the anti-Stokes radiation yield plateau is would be reached at $I \approx 1$ W/cm^2. However, in actuality the limiting value of D is markedly lower, mostly because of the existence of reverse (cross-relaxation) processes in the inter-ion relaxation of higher-energy states which govern the value of the denominator. Indeed, the probability or transfer coefficients for resonant energy transfer processes in the forward and back directions are equal. Therefore for the aforementioned values of $P \approx Q = 5 \cdot 10^{-15} cm^3 \cdot s^{-1}$, $C_A/C_0 \approx 0.2$ and $\tau_2 \approx 10^{-4}$(i. e., an order of magnitude less than for the one-photon state) we get a denominator for $D \sim 2 \cdot 10^3$, and consequently the yield plateau is reached only for $I > 10^3$ W/cm^2. This is why it is exceptionally important that the probabilities of forward (cooperative) and back (inter-ion cross-relaxation) energy transfer processes not be equal in cooperative luminophors. In the three-level model this can only be achieved by mismatch in the energy of electronic transitions, which should compensate for emission (or absorption) of lattice phonons in energy transfer processes.

T. Miyakawa and D. L. Dexter have proposed a theory [11,12] for such non-resonant energy transfer, based on a quasi-adiabatic approximation of multiphonon relaxation which they developed. The probability of this relaxation w_Φ for the weak electron-phonon interaction found in RE^{3+} ions may, in accordance with experiments [55-58], be described by an exponential:

$$w_\Phi(\Delta E) = w_\Phi(0) \exp(-\alpha \Delta E), \tag{32}$$

where $w_\Phi(\Delta E)$ is the multiphonon relaxation probability for width of the energy gap ΔE, $w_\Phi(0)$ is the limiting value of $w_\Phi(\Delta E)$ as $E \rightarrow 0$, and the coefficient α is equal to

$$\alpha = (\hbar \omega_\Phi)^{-1} \left\{ \ln \left[\frac{N_\Phi}{g} (\bar{n} + 1) \right] - 1 \right\}, \tag{33}$$

where $\hbar \omega_\Phi$ is the energy of lattice phonons; $N_\Phi = \dfrac{\Delta E}{\hbar \omega_\Phi}$ is the number of phonons needed to span the energy gap; $\bar{n} = \dfrac{1}{e^{\frac{\hbar \omega_\Phi}{kT}} - 1}$ is the number of popula-

ted phonon states at a given temperature; and g is the electron-phonon interaction constant. The corresponding probability (or coefficient) for energy transfer with a mismatch covered by the emission of N_Φ phonons is given by a similar exponential formula: ·

$$P(\Delta E) = P(0) \exp (-\beta \Delta E), \tag{34}$$

where $P(0) = \lim\limits_{\Delta E \to 0} P(\Delta E)$,and

$$\beta = \alpha - (\hbar \omega_\Phi)^{-1} \ln(1 + g_B/g_a). \tag{35}$$

Here the values g_a and g_B are the constants g for donor and acceptor ions of energy such that when $g_a \approx g_B$, the constants α and β are very simply connected:

$$\beta = \alpha - (\hbar\omega_\Phi)^{-1}\ln 2, \qquad (35)$$

from which

$$\beta = (\hbar\omega_\Phi)^{-1}\left\{\ln\left[\frac{N_\Phi}{g}(\bar{n}+1)\right] - 1 - \ln 2\right\}$$

for the case of phonon absorption.

This theory has been utilized by a number of authors for reporting experimental data and empirical determinations of the coefficients α and β, as well as values for $w_\Phi(0)$ and $P(0)$ for a number of RE^{3+} ions in various matrices. On the whole, they are in fair agreement with experiment although the quantitative values for α, β, $w_\Phi(0)$ and $P(0)$ organized in Table 2 still do not agree satisfactorily with one another. Various authors have treated $\hbar\omega$ somewhat differently, (as a limiting, most probable, average or lattice-effective value for the phonon energy), and have presented the Miyakawa-Dexter theory in a variety of different formulations which must lead to different forms of the constants α and β. In particular, in articles by Sergeyev [27-31] on calculation of the temperature dependence of cooperative luminescence yield, they have the form

$$w_\Phi(\Delta E, T) = w_\Phi(0, 0)\, e^{-\alpha\Delta E}\left\{\frac{(\bar{n}+1)^N}{\bar{n}^N}\right\} \qquad (37a)$$

and

$$P(\Delta E, T) = P(0, 0)\, e^{-\beta\Delta E}\left\{\frac{(\bar{n}+1)^N}{\bar{n}^N}\right\}. \qquad (37b)$$

The upper and lower multiplying factors in the curly brackets represent, respectively, phonon emission and absorption. In this way the constants α and β are determined for $T \to 0$ and it is postulated that the entire dependence on temperature of $w_\Phi(T)$ is imparted by that last factor, which was written down by taking the lattice photon spectrum into account. In an article by F. Auzel [59] an estimate was made of the probabilty of an optical absorption with the participation of phonons, and a somewhat different form was given for writing the coefficient α:

$$\alpha = (\hbar\omega_M)^{-1}\left[(1 - 2N)\ln\frac{N}{g} - 1 - \ln(\bar{n}+1)\right], \qquad (38)$$

and it was proposed to estimate this numerically through a single parameter — the maximum phonon frequency ω_M:

$$\alpha(\hbar\omega_M) = 5\cdot10^{-2}\exp(-6.5\cdot10^{-3}\hbar\omega_M) \qquad (39)$$

and

$$\beta = \alpha - (\hbar\omega_M)^{-2}\ln 2 \ , \qquad (39a)$$

where $\hbar\omega_M$ is given in inverse centimeters, α and β in centimeters.

In studies by Soviet authors (S. I. Pekar, N. M. Kristofel', K. K. Rebane and Yu. V. Perlin) which did not limit themselves to the case of weak electron-phonon interactions, formulas for estimating $w_\Phi(T)$ taking into account several lattice parameters were significantly more complicated [62,63]. One would think that, from the exponential dependence, this would lead to a very dramatic difference in the limiting efficiency of anti-Stokes luminescence. However, this does not happen because in the present case it is the ratio of P and Q which is important, not their absolute values. If there is some excess of energy ΔE significantly higher than kT in the forward direction (i. e., during cooperation), and there is the same deficit energy in the back direction (i. e., during the cross-relaxation decay of a two-photon excited state), then it is easy to show that the ratio

$$\frac{P}{Q} = \left(\frac{\bar{n}+1}{\bar{n}}\right)^N \approx e^{\frac{\Delta E}{kT}} \tag{40}$$

and does not depend on the sundry coefficients that determine the value of P and Q. For $\Delta E \approx 1000$ cm^{-1}, obtained in the presence of sensitizers, and $kT \approx 210$ cm^{-1} (corresponding to room temperature), we get $\exp(\frac{\Delta E}{kT}) \approx 10^2$ to a first approximation. In this manner,

Table 2. Constants Determining the Probability of Multiphonon Relaxation and Phonon-Assisted Energy Transfer for RE^{3+} Ions in Various Bases

Base	Constant of Multiphonon Relaxation		Energy of Lattice Phonons		Constant of Energy Transfer	Electron-Phonon Interaction $P(0,0)$cm^{3}s^{-1}		Constant of Electron-Phonon Interaction	Reference
	α, cm	$w_\Phi(0)$	limiting	predominant	β, cm	UV	IR	g	
Flourides									
YF$_3$	$5\cdot10^{-3}$	10^8	375	—	$2,5\cdot10^{-3}$	10^{-17}	—	0,02—0,03	[26]
	$5\cdot10^{-3}$	—	—	190—270	$4\cdot10^{-3}$	—	—	—	[27]
	$5,5\cdot10^{-3}$	—	—	—	$2,4\cdot10^{-3}$	—	—	—	[59]
LaF$_3$	$5\cdot10^{-3}$	10^8	375	350	$3\cdot10^{-3}$	$2\cdot10^{-17}$	$5\cdot10^{-17}$	0,05	[11]
Oxides									
Y$_2$O$_3$	$3,8\cdot10^{-3}$	10^8	550	430	$(2,2—2,5)\cdot10^{-3}$	10^{-16}	—	—	[58]
	$5,1\cdot10^{-3}$	—	430	—	—	—	—	—	[60]
Oxyhalogenides									
YOCl	$3\cdot10^{-3}$	—	—	110—500	$1,5\cdot10^{-3}$	—	—	—	[27]
YOF	$2\cdot10^{-3}$	10^9	620	570	10^{-3}	10^{-16}	—	0,07	[59]
Y$_3$OCl$_7$	$4,5\cdot10^{-3}$	$3\cdot10^9$	620	550	—	$4\cdot10^{-16}$	—	—	[26]
	$2\cdot10^{-3}$	—	—	600	10^{-3}	—	$5\cdot10^{-16}$	0,07	[59]

The dashes represent a lack of data

25

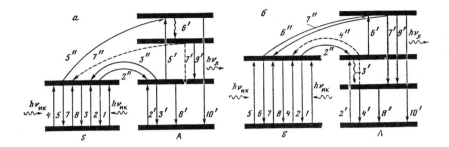

Fig. 11. Scheme for Converting IR to Visible in a Four-Level Model of Activator Ions (*A*) With Sequential Sensitization by Other Ions (*S*)

a- Multiphonon relaxation to an intermediate level from a two-photon (doubly-pumped) ion state; A,b - Multiphonon relaxation to an intermediate level from a one-photon ion state A. The numbers denote the sequence of the transitions, identical numbers correspond to processes going on simultaneously. The width of the horizontal lines symbolizes the possible splitting of the corresponding level.

$$\frac{\eta}{[1-(2+\gamma\tau_2^*)\eta]^2} \approx e^{\frac{\Delta E}{kT}} \frac{\tau_1}{\tau_2^*}(\sigma_A\tau_1)\,I,\tag{41}$$

where I is in photons$\cdot$cm$^{-2}\cdot$s^{-1}.

Further on we will make this approximate value more precise by including the real energy structure of Yb and Er ions for certain crystal bases and the two fundamental emission bands of Er ions.

2.3. Four-level Model of Radiating Centers.

The four-level model of radiating ions in cooperative luminophors has distinct advantages over the three-level model, as is the case with centers in optical quantum generators (lasers)[51,64]. For anti-Stokes luminescence, two fundamental types of four-level systems are possible, namely: 1) the transition to a secondary lower-lying energy level, responsible for the emission of a visible-light photon, is accomplished after accumulation of the energy of two excitations, 2) the transition to a level of lower energy proceeds from an intermediate one-photon state before the summing of pumping energy (Fig. 11a, b).

The first model is better suited for an explanation of the Er^{3+} green emission, insofar as energy summing usually leads to filling the $^2H_{11/2}$ and even $^4F_{7/2}$ states, then a radiative transition to the ground state takes place after intra-center relaxation to $^4S_{3/2}$. The second model is more likely for the Er^{3+} red emission ($^4F_{9/2}\to{}^4I_{15/2}$ transition) taking into account the intermediate relaxation from the one-photon state ($^4I_{11/2}\to{}^4I_{13/2}$). In addition to this, a red band is possible, especially during short wavelength pumping [45,19], as a result of the multiphonon relaxation of $^4S_{3/2}\to{}^4F_{9/2}$, i. e. as per the first model or some other more complicated channel [2,3]. In all these cases we apparently have additional Stokes energy losses ($h\nu_{\text{изл}}<2h\nu_{\text{ик}}$), but the gain in quantum yield can more than make up for it.

In fact, the obvious advantage of the first model consists of the fact that it is possible to improve the relationship between summing and decay

26

processes in the two-photon state by creating an energy trap ("hole") in that state which requires the absorption of several phonons for decay. This situation is realized to some degree in the Er^{3+} green emission band, especially in the presence of a sensitizer. We will however not make a special case of this, since even under conditions of $w_\Phi \gg QC$ this model is practically identical to the three-level model except for different back-transfer constants. This advantage is not so great for Er^{3+} ions, because the ${}^4S_{3/2}$ state can decay into two ${}^4I_{13/2}$ states with the emission of a lattice phonon, and that is significantly more likely given identical absolute values of energy mismatch.

The picture is much more complicated for particle relaxation of excitation. In this case the kinetic equations of balance in the absence of a sensitizer takes on the form

$$\frac{dN_1}{dt} = \sigma_A (C_A - N_1 - N_1' - N_2) I + Q_A N_2 (C_A - N_1 - N_1' - N_2) -$$
$$- \gamma_1 N_1 - N_1/\tau_1 - P_A N_1 N_1' + \gamma_2 N_2 = 0,$$

$$\frac{dN_1'}{dt} = \gamma_1 N_1 - N_1'/\tau_1' - P_A N_1 N_1' + Q_A N_2 (C_A - N_1 - N_1' - N_2) = 0, \tag{42}$$

$$\frac{dN_2}{dt} = P_A N_1 N_1' - N_2/\tau_2 - Q_A N_2 (C_A - N_1 - N_1' - N_2) = 0.$$

Here N_1', τ_2' and γ are the concentration, lifetime and probability of relaxation to an intermediate level, P_A and Q_A are coefficients of cooperation and cross-relaxation in the summing of excitations, and the rest have the same meaning as before.

Neglecting the effect of ground state depletion (cf. section 2.2) and combining all three equations (after multiplying the bottom one by a factor of 2), as before, we obtain the overall balance of dissipative transitions in the system:

$$\frac{N_1}{\tau_1} + \frac{N_1'}{\tau_1'} + 2\frac{N_2}{\tau_2} = \sigma_A C_A I + \gamma_2 N_2, \tag{43}$$

or, with consideration of Eq. (42) and the anti-Stokes luminescence quantum yield from Eq. (9), (in the present case $\eta = \dfrac{N_2/\tau_*^2}{\sigma_A (C_A - N_1 - N_1' - N_2) I}$), we obtain

$$\frac{N_1}{\tau_1} + \frac{N_1'}{\tau_1'} = \sigma_A C_A I [1 - \eta (2 - \gamma_2 \tau_2^*)]. \tag{44}$$

On the other hand, N_1' and N_1 are connected by

$$N_1' = \left(\gamma_1 N_1 - \sigma_A C_A I \eta \frac{\tau_2^*}{\tau_2} \right) \tau_1'. \tag{45}$$

Substituting Eqs. (45) and (42) into (44) we find N_1, and likewise N_1' from (45):

$$N_1 = \frac{\sigma_A C_A I (1 - \eta) \tau_1}{1 + \gamma_1 \tau_1},$$

$$N_1' = \frac{\sigma_A C_A I \tau_1'}{1 + (\gamma_1 \tau_1)^{-1}} \left[1 - \eta \left(1 + \frac{\tau_2^*}{\tau_2} + \frac{1}{\gamma_1 \tau_1} \frac{\tau_2^*}{\tau_2} \right) \right]. \tag{46}$$

All this allows us to determine the quantum yield of anti-Stokes luminescence. From the bottom equation in (42) we have

$$\eta = \frac{P_A N_1 N_1'}{1 + Q_A C_A \tau_2} \frac{\tau_2}{\tau_2^*} \frac{1}{\sigma_A C_A I} . \tag{47}$$

Substituting N_1 and N_1' from Eq. (46) into this, we obtain a final expression for η in the following form:

$$\frac{\eta}{(1-\eta)\left[1 - \eta\left(1 + \frac{\tau_2^*}{\tau_2} + \frac{1}{\gamma_1 \tau_1} \frac{\tau_2^*}{\tau_2}\right)\right]} = \frac{P_A \tau_1 C_A \sigma_A \tau_1' \frac{\tau_2}{\tau_2^*} I}{(1 + Q_A C_A \tau_2)(1 + \gamma_1 \tau_1)\left(1 + \frac{1}{\gamma_1 \tau_1}\right)} , \tag{48}$$

similar to Eq. (25) for the three-level system. All the conclusions we made earlier still hold for this case, but there are some differences caused by the introduction of intermediate levels.

The advantage of the given system is that we can use longer-lived states with $\tau_1' > \tau_1$, while the effective pumping radiation absorption cross-section σ_A corresponds to the lifetime τ_1 ($\sigma_A \tau_1 = \text{const}$). At the same time, there are additional terms appearing in the denominator of the right-hand side for which it is easily shown that maximum yield at small I corresponds to $\gamma_1 \tau_1 \approx 1$, i. e. to the case in which the probability of a radiative transition of the ion from a one-photon state to the ground state and its radiative/multiphonon transition to a longer-lived state are equal. As such, there appears a multiplying factor of 1/4 on the right-hand side of Eq. (48), so that it is difficult to realize this advantage in systems with Er^{3+}, where τ_2' is only a few times larger than τ_1. Moreover, at high pumping densities the limiting value of radiation yield is somewhat lower than in the three-level system, where $\eta_M = 33\%$. Indeed, assuming $\gamma_1 \tau_1 = 1$ and $\gamma_2 \tau_2^* = 1$ in Eq. (48) we get for high I, $\eta_M \leq 20\%$, which is fairly close to the maximum efficiency of the red anti-Stokes emission obtained during laser pumping at high powers. It turns out that it is also difficult to realize the condition $P_A \gg Q_A$ in this system because during the formation of an energy trap (the energy deficit for its decay to the $^4I_{11/2}$ and $^4I_{13/2}$ states) in the two-photon (excited) state, its decay will take place with emission, not absorption, of a phonon and the formation of two intermediate states ($^4I_{13/2}$) with identical lower energies will result.

The accompanying advantageous features of the four-level model are more fully revealed when we introduce a sensitzer having no states of lower energy than the one-photon state of the activator. Herein lies the unique property of Yb^{3+} and, at the same time, the fundamental reason for why the red emission band of Er^{3+} obtains a higher anti-Stokes conversion efficiency than the green. In a similar positive way the splitting of the Yb^{3+} ground state has a rather large impact on the value of η (~ 700 cm^{-1}). It is because of this that at room temperature and lower it can give up a smaller portion of energy (during summing of excitations) than it transfers (during the decay of a two-photon excitation).

An examination, analogous to that previously put forth, of a sensitized four-level system with highly incoherent interactions between activators and sensitizers in intermediate states results in the following expression for η:

$$\frac{\eta}{(1-\eta)\left\{1-\eta\left[1+\frac{\tau_2^*}{\tau_2}\left(1+\frac{1}{\gamma_1\tau_1}+\frac{P_{10}}{P_{01}}\frac{C_S}{C_A}\frac{1}{\gamma_1\tau_S}\right)\right]\right\}}=$$

$$=\gamma_1\tau_1'\frac{P_{01}}{P_{10}}\frac{C_A}{C_S}\frac{\tau_2}{\tau_2^*}P_S\sigma_S\tau_S^2 I\left[1+(Q_SC_S+Q_AC_A)\right]^{-1}\times$$

$$\times\left[1+\frac{P_{01}}{P_{10}}\frac{C_A}{C_S}\left(\frac{\tau_S}{\tau_1}+\gamma_1\tau_1'\right)\right]^{-2}. \qquad (49)$$

To understand the physical meaning behind this expression without having previously examined the much simpler cases would have obviously been very difficult. As it is, based on the previous results (48) obtained for the sensitized three-level system and the four-level system without sensitizers, it is easy to show that the primary advantage of sensitizers in the present case is only that they serve to increase the ratios

$$\frac{P_SC_S\tau}{1+(Q_SC_S+Q_AC_A)\tau_2} \quad \text{and} \quad \frac{P_AC_A\tau_1}{1+Q_AC_A\tau_2}.$$

Meanwhile, other terms remain constant ($\sigma\tau$) or, conversely, even serve to decrease the gain from the introduction of sensitizers. Thus, as was the case in the three-level model, there exist several different optimal ratios of activator and sensitizer concentrations for which the yield is maximum. However, the yield tends toward a constant value as the activator and sensitizer concentrations are increased (up to $C_S+C_A=C_0$), requiring us to introduce the complicating conditions which we examined earlier.

We will dwell briefly on the question, which is important in a practical regard, of the effect of the base lattice on the ratio of intensities of the green and red anti-Stokes emission bands in Yb^{3+} and Er^{3+} ions generated by the foregoing scheme (Fig. 11a, b). An empirical rule is used in the literature [65] which holds that the green band is predominant if the phonon frequency is limited to $\hbar\omega_\Phi>500$ cm^{-1}. A change in the probability of intra-center multiphonon relaxation by $^4I_{11/2}\rightarrow{}^4I_{13/2}$ is indicated to be the primary reason for the influence of $\hbar\omega_\Phi$ on the luminescence light [56,58]. Such intra-center relaxation also takes place radiatively (i. e., even in the absence of electron-phonon interactions), as evidenced by, for example, the developement of IR lasers based on this transition [66,67] with output in the 3 μm range. Therefore in our study [68] we proposed and justified a new supposition, that the primary reason for this influence is a change in the probabilities of inter-ion transitions, not intra-center, i. e. in the probabilities of cooperation of energy via both channels taking into account phonon energy and the transition energy mismatch in two-photon states.

2.4. Summing of Three or More Electronic Excitations. In a series of experiments [1-3,9,10,70-74], with pumping in the 1.5 μm range corresponding to absorption by Er^{3+} ions (the $^4I_{15/2}\rightarrow{}^4I_{13/2}$ transition), red, green, and even indigo emission bands were observed. Indigo and ultraviolet bands from Er^{3+} ions arose also during pumping at 0.9 to 1.0 μm [2,3,75]. From simple energy considerations it is understood that the generation of anti-Stokes luminescence in such cases requires the summing of three or more electronic excitations, not just two. The Bloembergen model (sequential absorption) was used to explain these effects in the initial experiments [9,10], after which, in the article by V. V. Ovsyankin and P. P. Feofilov [69], a model was proposed of triply cooperative sensitization (at least for the green band of Er^{3+} at

29

$\lambda_B = 1.5$ μm and the indigo band of Tm^{3+} at $\lambda_B = 0.8$ to 1.0 μm). In addition, our studies [71-74] demonstrated that the significantly more efficient red emission by Er^{3+} ions at $\lambda_B = 1.5$ μm arose via the mechanism of sequential sensitization. During this we discovered a new effect, that of a substantial increase in the intensity of anti-Stokes luminescence by doping with sensitizers (Yb^{3+}) that do not absorb nor change the absorption of Er^{3+} ions in that range of the spectrum. The mechanism of sequential sensitization was favored by a number of other authors [70,32] for interpreting the various effects in summing three or more electronic excitations.

We will investigate the kinetic limits to the efficiency of anti-Stokes luminescence within the framework of just such a model, using it to obtain higher radiation yield at a given IR pumping density. The corresponding highly-simplified system must include, of course, four energy levels in the participating ions, which are spaced equidistantly. At the same time, the possibilities for such a model are fairly broad and it may be used for a first approximation description of anti-Stokes luminescence during particle intra-center Stokes relaxation of the first and subsequent electronic excitations (for example, the second one for the red band in Er^{3+}), when the energy of a visible-light photon becomes markedly lower than the triple energy of the initial excitation. With the help of this model we may qualitatively, and somewhat quantitatively, explain the effect of sensitizers on the increase in radiation efficiency caused by a change in the probabilities of intermediate processes of summing and relaxing electronic excitations.

In light of the previous assumption about the distribution of an excitation in the ion system and its uniformity, the corresponding kinetic equations have the form

$$\sigma_A C_A I - \frac{N_1}{\tau_1} - 2P_1 N_1^2 + 2Q_1 N_2 C_A - P_2 N_2 N_1 + Q_2 N_3 C_A + \gamma_2 N_2 = 0,$$

$$P_1 N_1^2 - Q_1 N_2 C_A - \frac{N_2}{\tau_2^*} - \gamma_2 N_2 - P_2 N_1 N_2 + Q_2 N_3 C_A + \gamma_3 N_3 = 0, \qquad (50)$$

$$P_2 N_1 N_2 - Q_2 N_3 C_A - \frac{N_3}{\tau_3^*} - \gamma_3 N_3 = 0,$$

which uses previous conventions as well as the following (cf. Fig. 11): N_3 is the concentration of excited particles in the final radiative state, τ_3^* is the inverse probability for radiative transition from this state to the ground state, γ_3 is the probability for relaxation in state 2; and P_2 and Q_2 are the coefficients for cooperation and cross-relaxation. We note that τ_3, τ_3^* and γ_3 are connected in the following way:

$$\frac{1}{\tau_3} = \frac{1}{\tau_3^*} + \gamma_3 \qquad (51)$$

Combining these equations (the second is multiplied by 2, the third by 3), we obtain the balance of the dissipative transitions in the investigated system:

$$\frac{N_1}{\tau_1} + 2\frac{N_2}{\tau_2^*} + N_2\gamma_2 + 3\frac{N_3}{\tau_3^*} + N_3\gamma_3 = \sigma_A C_A I \equiv \Omega. \qquad (52)$$

Or, by taking the quantum yield into account

$$\eta = \frac{N_3}{\tau_3^*}\frac{1}{\Omega} = \frac{P_2 N_1 N_2 \tau_3/\tau_3^*}{1 + Q_2 C_A \tau_3}\frac{1}{\Omega} \qquad (53)$$

30

in the form

$$\frac{N_1}{\tau_1} + 2\frac{N_2}{\tau_2^*}\left(1 + \frac{\gamma_2\tau_2^*}{2}\right) = \Omega\left[1 - 3\eta\left(1 + \frac{1}{3}\gamma_3\tau_3^*\right)\right].\qquad(54)$$

From the second Eq. (50), N_2 is expressed in terms of N_1 as

$$N_2 = \frac{(P_1 N_1^2 - \Omega\eta)\,\tau_2^*}{1 + \gamma_2\tau_2^* + Q_1 C_A \tau_2^*}.\qquad(55)$$

In this way, taking the ratio N_3/τ_3^* from the last Eq. (50) and taking advantage of the relation (53), the value of η may be described by

$$\eta = \frac{P_2 N_1 N_2/\Omega}{1 + Q_2 C_A \tau_3}\frac{\tau_3}{\tau_3^*} = \frac{P_2 N_1 (P_1 N_1 - \Omega\eta)\,\tau_2^*}{(1 + \gamma_2\tau_2^* + Q_1 C_A \tau_2^*)(1 + Q_2 C_A \tau_3)}\frac{\tau_3}{\tau_3^*}\frac{1}{\Omega},\qquad(56)$$

where N_1 is determined by solving the rather unwieldy complete quadratic equation:

$$N_1^2 + \frac{1 + \gamma_2\tau_2^* + Q_1 C_A \tau_2^*}{P_1\tau_1\,(^1/_2 + \gamma_2\tau_2^*)}\,N_1 - \frac{\Omega}{P_1}\cdot\frac{1 + \gamma_2\tau_2^* + Q_1 C_A \tau_2^*}{^1/_2 + \gamma_2\tau_2^*}\,(1 - 3\eta\theta) = 0\qquad(57)$$

for

$$\theta = 1 + \frac{1}{3}[\gamma_3\tau_3^* - (2 + \gamma_2\tau_2^*)\cdot(1 + \gamma_2\tau_2^* + Q_1 C_A \tau_2^*)^{-1}].$$

Since an analysis of the general expression for η is so difficult, we will carry it out here for two limiting cases, and then in the following review of the experimental data we will employ a numerical solution.

At low pumping densities, when the first stage of cooperation is still of a low probability $(P_1 N_1 \ll \frac{1}{\tau})$, the value of $N_1 = \Omega(1 - 3\eta\theta)\tau_1$. Since N_2 is positive over this regime, $\Omega\eta \lesssim P_1 N_1^2 \ll \Omega(1 - 3\eta\theta)$, i. e. $\eta \ll 1$, and from this we clearly cannot expect high efficiency from anti-Stokes luminescence caused by the summing of three excitations.

In this case the radiation yield is proportional to the square of the pumping intensity:

$$\eta = \frac{P_2 P_1 \tau_1 \tau_2^* C_A^2\,(\sigma_A\tau_1)^2\,I^2}{[1 + (\gamma_3 + Q_2 C_A)\,\tau_3^*]\,[1 + (\gamma_2 + Q_1 C_A)\,\tau_2^*]}.\qquad(58)$$

Here as before, in finding the optimal pumping in the maximum IR absorption band and with a radiative transition probability τ_1^{-1}, the value of $I_0 = (\sigma\tau)^{-1}$ depends not on the forbiddenness or multipole order of the transition, but only on the spectral width of the band and the wavelength of its maximum (the latter at $\lambda = 1.5$ μm leads to about a twofold decrease of I_0 compared to I_0 for $\lambda = 1.0$ μm).

This erases most of the advantage gained from a longer lifetime of the first excited state of Er^{3+} $(^4I_{13/2})$, which is 5 to 10 ms, in that the radiation efficiency is not dependent on it. For moderate concentrations C_A and small γ_0 and γ_3, and when the decay of two- and three-photon states as determined by the value of the denominator in Eq. (58) may be ignored, then the value of η is proportional to the product of the coefficients of excitation summing

31

and the intermediate states lifetimes, and to the square of the activator concentration. For the red band of YOCl:Er luminophors (corresponding to the $^4F_{9/2} \to {}^4I_{15/2}$ transition) we obtain a maximum value for this product of $2 \cdot 10^6$, so that, within the assumptions that we have made, we can expect radiation yield $\eta \approx 0.2\%$ by the time $I \approx 0.1$ W/cm^2.

This value for η is less than the experimental value [36,41,43] by about two orders of magnitude for single-activator YOCl:Er luminophors (10%) and by a little less than one order of magnitude for Yb doping (10%). Naturally these discrepancies can be explained by the existence of other decay channels for the double and triple excitations (denominator $\gg 1$), primarily cross-relaxation and multiphonon channels. Taking these processes into account leads to surprising agreement with experimental data. As with the summing of two excitations, doping with Yb^{3+} ions, which reduces the intermediate excited state lifetime, still and all increases the overall radiation efficiency because of the increase in the ratio P_2/Q_2. In the present case this sensitization method operates exclusively, since the absorption of pumping radiation does not at all depend on the presence of Yb.

The efficiency of anti-Stokes luminescence is not so critically dependent on the value of P_1 also at higher pumping densities, when the dependence of the luminescence brightness on I starts to change from cubic (at $I > 0.1$ W/cm^2). Actually, in concordance with Eq. (57) for $P_1 N_1 \gg 1/\tau$ the values for N_1 and N_2 are determined to be

$$N_1 = \left[\frac{\Omega}{P_1} \frac{1 + \gamma_2 \tau_2^* + Q_1 C_A \tau_2^*}{1/2 + \gamma_2 \tau_2^*} (1 - 3\eta\theta) \right]^{1/2},$$

$$N_2 = \frac{\Omega (1 - 3\eta\theta) \tau_2^*}{2 [1 + (\gamma_2 + Q_1 C_A) \tau_2^*]}. \tag{59}$$

In the same way we obtain for η

$$\frac{\eta}{(1 - 3\eta\theta)^{1/2}} = \frac{P_2 (\sigma_A C_A I)^{1/2}}{P_1^{1/2}} \frac{(1 + \gamma_2 \tau_2 + Q_1 C_A \tau_2^*)^{1/2}}{(2 + \gamma_2 \tau_2)^{1/2}} \frac{\tau_2^*}{1 + Q_2 C_A \tau_3^*} \frac{\tau_3}{\tau_3^*}, \tag{60}$$

from which, from the condition that the denominator on the left-hand side of this equation be positive, we obtain

$$\eta_M < \frac{1}{3 + \gamma_3 \tau_3^*}. \tag{61}$$

This agrees with the experimental data, in that the radiation yield will grow as $I^{1/2}$ since it is inversely proportional to $P_1^{1/2}$.

This somewhat unexpected $\eta(I)$ dependence arises because anti-Stokes luminescence is forced to compete with similar anti-Stokes transitions in, as a matter of fact, the IR region (an interesting species of nonlinear quenching) which at the same time contribute to the emissions that we see. Thus a significant departure from cubic dependence of luminescence intensity J_1 on I begins long before the anti-Stokes luminescence yield plateau ($\eta \approx 1\%$) is reached, and even before the onset of phototropism of the sample relative to the pumping radiation. In this regime, as during lower pumping densities, doping with Yb^{3+} ions increased the luminescence yield; the estimate of it from Eq.(60) is in satisfactory agreement with our experiment. It is necessary to consider, then, that in view of the interaction of center parameters (σ_A and τ_1, P_1 and τ_1), transitions from one regime to another determined by $P_1 N_1 > 1/\tau_1$ arise for $I = I^* = (P_1 \tau_1 C_A)/(\sigma_A \tau_1)$, i.e. they depend not on the multipole order or allowedness of the transition, but strongly on λ (λ^8 for dipole-dipole and λ^{12} for quadrupole-quadrupole transitions), as well as on the spectral width of the transition.

This competition between anti-Stokes transitions of different orders obviously should occur during the summing of four or more single excitations (at present, no emission band has been discovered which could only be interpreted as the cooperation of more than four excitations, but we will not exclude the possibility that such a band might be discovered in the future).

It is interesting to note that the bands generated during the summing of four excitations, in particular the blue band of Er^{3+} (the $^2H_{9/2} \to {}^4I_{15/2}$ transition), can be generated with almost the same probability as the result of two double excitations ($^4I_{9/2}$) or one triple and one single excitation. Proceeding from the examination conducted above on summing for the output of anti-Stokes luminescence, it is clear that the dependence $\eta\varphi(\eta) \sim I^3$ will be observed only until pumping intensity $I < I^*$. A much weaker dependence of $\eta\varphi(\eta) \sim I$ obtains for large I, when the lower-lying excited states are depopulated to the ground state because of their pair-wise summing for both models of the radiation following summing of four excitations. Therefore reaching the plateau in anti-Stokes luminescence yield η will in this case also occur very gradually, and this is in agreement with the results of our experiments.

Since Yb^{3+} doping increases the intensity of the red band of Er^{3+} but simultaneously exitinguishes the green and indigo bands, then by all appearances it is the summing of two "double" excitations that is the most probable in YOCl:Yb,Er luminophors. In the final analysis, the dominance of one or another channel for the summing of four single excitations is determined by the relation

$$\frac{P(E_0, E_0)\, P(2E_0, 2E_0)}{P(E_0, 2E_0)\, P(E_0, 3E_0)} \cdot \frac{(1/\tau_3 + Q_2 C_A)}{(1/\tau_2 + Q_2 C_A)}, \tag{62}$$

where the energy of the summed states is the argument, which depends on the ratio of the corresponding coefficients of cooperation as well as on the time for intra-center relaxation of excitations. Therefore, conversely, summing processes of single and triple one-photon excitations may, in principle, be dominant in other bases.

3. SPECTRAL, ENERGY AND KINETIC PROPERTIES OF COOPERATIVE LUMINOPHORS DOPED WITH Yb^{3+} AND Er^{3+} IONS

We will examine the fundamental results of experimental investigations into the spectral, energy and inertial (delay) characteristics of both anti-Stokes and Stokes luminescence in cooperative luminophors of various polycrystalline bases doped with Er^{3+} and Yb^{3+} ions and compare them with the data found in the literature, as well as to the results of analysis of the idealized model of luminophors presented in the previous section.

33

Apart from the exposition of results from our original research, our aim is also to screen the literature for the most reliable data on energetic structure and intra-center/inter-ion transition probabilities for Er^{3+} and Yb^{3+} ions in various crystal bases, which is needed for comparison of theory with experiment. The final goal of such comparison is to clarify the possibilities for quantitative analysis of the energetic structure of anti-Stokes luminescence based on the contemporary theory of RE^{3+} ion electronic excitation cooperation and relaxation, as well as refining the outlook for further improvement in this class of luminophors, which is vital in the developement of practical applications.

3.1   Spectral Characteristics of Anti-Stokes Luminophors.   The spectral properties of luminophors, as we know, include the spectra of excitation, absorption (or diffuse reflection, which is easier to measure in polycrystalline samples) and emission. In cooperative luminophors, as with laser crystals activated with RE^{3+} ions, these spectra consist of relatively narrow structured bands, 100 to 1000 cm^{-1}, occupying the greater part of a fairly wide spectral region (from 0.3 to 3 μm). The weak effect of the crystal field and electron-phonon interaction on the location of the bands, which are caused by transitions inside the unfilled $4f$ shells of the RE^{3+} ions, makes the identification of optical transitions in various bases much simpler; broad gaps in these spectra are testimony to the high purity of the original materials, which contains additional impurities (including other RE^{3+} or RE^{2+} ions) in concentrations of no more than $1 \cdot 10^{-4}$ mol. %.

For practical purposes it is usually enough to know only the approximate location of bands of IR excitation and anti-Stokes visible radiation, and limitations were set on such determinations in the first articles on the subject [5-8], all the more since the highly linear dependence of luminescence intensity on IR pumping intensity makes high-spectral-resolution measurements more difficult. At the same time studying the mechanisms of processes of generation, transfer and relaxation of electronic excitations requires, of course, more detailed knowledge of the structure of all states that might be taking part in the various stages of development and quenching of anti-Stokes luminescence. Above all, an experimental classification is required for the characteristic IR absorption bands of activator and sensitizer ions, as well as determination of the term structure of the ground and first excited state of these ions.

In the first work on this subject [25,71,76-79], carried out on oxychloride and fluoride bases, it was established that the IR pumping spectra (in the 0.9 to 1.1 μm range) for exciting visible luminescence in various activators (Er^{3+}, Ho^{3+}, Tm^{3+}) doped with the same sensitizer (Yb^{3+}) were usually identical to each other. This means that, in the majority of cases for cooperative luminophors, sensitizer absorption predominates over activator absorption, i. e. this is so at least for two-step sensitization of the radiating ions. This conclusion is found to be in good agreement with the difference in one-photon excited state lifetimes of the activator and sensitizer (as much as a factor of 10) and with the usually higher concentration of sensitizers (from 5 to 50 times higher) in cooperative luminophors with maximum luminescence efficiency. However, in oxysulfide bases, for which the optimal concentrations C_{Yb} (8%) and C_{Er} (6%) are fairly close to one another, and the transition times are also commensurate, the absorption by Yb and Er is of the same order of magnitude, i. e. one-step sensitization of radiating ions directly excited by IR radiation may take place [44,79,80]. Even for YOCl bases, in which total sensitizer absorption exceeds activator absorption by more than an order of magnitude, it is possible to get selective excitation (in the 0.97 μm range) to directly excite just the activator ions (Fig. 12). Nonetheless, doping with sensitizers in this case increases Er^{3+} luminescence efficiency by almost an order of magnitude [81]. As the experiments directly confirm, this is because the action of the sensitizer is in no way restricted by increased absorption of pumping radiation (all the more so since the one-photon excited state lifetime falls off by just about the same factor, cf. section 2.2). This was exhibited in an especially graphic way in the sensitization effect (as much as a factor of 10), which we discovered [50,70,73] in the red emission band of Er^{3+} during pumping at a different IR band of Er^{3+} (1.5 to 1.6 μm), in which there was no absorption by Yb^{3+} whatsoever, and doping it in did not increase absorption by Er^{3+} ions (e.g., by weakening the forbiddenness of transitions).

In spite of the fact that the excitation spectra of anti-Stokes luminophors had been studied in a series of papers [25,59,80,81], the Stark structure of the ground ($^2F_{7/2}$) and first excited state ($^2F_{5/2}$) of Yb^{3+} ions for bases was completely determined for the first time only in [81]. In addition to IR excitation spectra of Yb^{3+} and Er^{3+}-doped luminophors (Figs. 12, 13 and 14), the IR excitation and emission spectra of YOCl bases doped only with Yb^{3+} ions over the temperature range 77 to 300 K (Fig. 15) was measured in that article. On the whole, the results of these measurements (carried out with an accuracy down to 10 cm^{-1}) (Table 3) were in satisfactory agreement with the results of [59], which earlier had found the Stark structure only of the excited state of Yb^{3+} (at 4.2 K), and also with the much later work [47] which determined that very structure for the crystallographically-analogous GdOCl base.

The fundamental discrepancy with the data in [59] is the absence of a 10,283 cm^{-1} component, probably arising due to Er^{3+} ion absorption, which takes place in precisely this region of the spectrum ($\lambda \approx 970$ nm). We note that the existence of four components for this state in Yb^{3+} contradicts crystal field theory. According to this theory, the crystal field causes Stark level splitting with a maximum number determined for half-integer j by the formula $j + \frac{1}{2}$ and for $j = \frac{5}{2}$ equal to 3. The discrepancies in the data for YOCl and GdOCl generally do not exceed 30 cm^{-1}, which lies within the range of shifts for other lattices of this same type (such as $Lu_3Al_5O_{12}$ and $Y_3Al_5O_{12}$ [51]). At the same time, there is a significant difference in the data for the upper Stark component (710 cm^{-1} in our article and 391 cm^{-1} in [47]). Apparently this is connected with the fact that Yb^{3+} radiation in the 1.066 μm range lies close to the sensitivity limit of the photoelectron multiplier used in [47], while in our paper we used for that purpose either germanium photodiodes or thermocouples with correspondingly constant quantum or energetic sensitivity over that region of the spectrum. This difference is of sizeable importance, inasmuch as it determines the minimum energy which can be transferred from Yb^{3+} to Er^{3+} ions in the second stage of energy summing. Besides this, it also defines the possibilities of using YOCl or GdOCl luminophors for recording radiation fields of YAG:Nd lasers. It is well-known [82] that a partial substitution of Yb for Gd leads to a two- and three-fold increase in sensitivity over this very spectral region, i. e. it increases the probability of the given transition (Gd ions absorb only in the near ultraviolet). The intensity of this longer-wave transition in luminophors

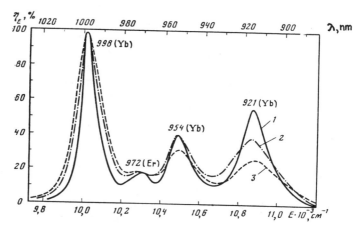

Fig. 12. IR Excitation Spectra for Visible Luminescence in an Oxychloride Base at Various Temperatures

1- 77 K; 2- 293 K; 3- 370 K. The numbers at the maxima correspond to the wavelength in nanometers.

Fig. 13. IR Excitation Spectra for Anti-Stokes Luminescence in BaYF$_5$:Yb, Er (1), BaYF$_5$:Yb,Ho(2) and YOCl:Yb,Er (3) and Their Agreement With the Emission Spectrum of GaAs:Si (Д)

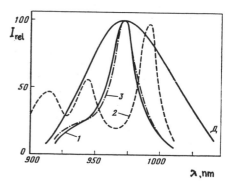

Fig. 14. Fine Structure of the Excitation Spectrum of YOCl:Yb,Er Luminophor at Various Temperatures

Fig. 15. Spectrum of an Incandescent Lamp (0.9 to 1.1 μm) After Passing Through a Compressed Layer at 77 K of Luminophors YOCl:Yb,Er (a) and YOCl:Er (b). (Narrow lines appear over the continuous background for both absorption and radiation of ions).

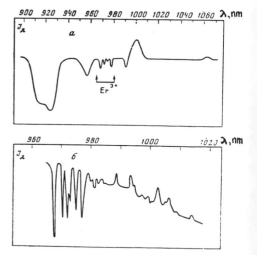

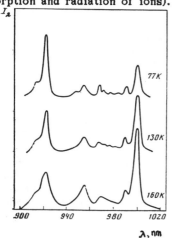

36

Table 3. Stark Components of Yb^{3+} Ion Terms for Oxychloride Bases (cm^{-1})

Терм	YOCl [81]	YOCl [59]	Y_3OCl_7 [59]	GdOCl [47]
$^2E_{7/2}$	0	0	0	0
	94	—	—	94
	195	—	—	166
	710	—	—	391
$^2F_{5/2}$	10 098	10 090	10 219	10080
	10 565	10 283	10 520	10596
	10 951	10 572	10 700	10962
		10 837		

The dash represents a lack of data.

without Gd is roughly 4% of the intensity of a transition at 1.0 μm, which is fundamental in the excitation spectrum of these luminophors at room temperatures.

The IR excitation spectra of other bases, primarily fluorides and oxysulfides (Figs. 16-18) which occupy second place as regards absolute anti-Stokes luminescence efficiency, differs somewhat from the oxychloride spectrum found at 300 K in the region around 0.91 to 1.02 μm. It is markedly narrower and grouped basically in the 0.94 to 0.98 μm region (10,640 to 10,200 cm^{-1}); the IR luminescence spectrum is correspondingly narrower, located from 0.98 to 1.02 μm (10,200 to 9800 cm^{-1}). This testifies to the fact that the overall Stark splitting of the ground and excited state of Yb^{3+} is about the same in fluoride and oxysulfide bases, being about 400 to 500 cm^{-1}. Fluoride bases have an even narrower excitation spectrum (in particular, $NaYF_4$ [83-85] and $BaYF_5$, as well as LaF_3), which is dominated at $T=293$ K by a naturally-occuring sharply-defined maximum in the 970 to 980 nm region (10,200 to 10,300 cm^{-1}). This peak rather closely coincides ($\Delta E < 100$ cm^{-1}) with the location of the lower Stark component of the $^4I_{11/2}$ state of Er^{3+} ions, which was determined earlier primarily in studies on laser crystals (LaF_3, $LiYF_4$, BaY_2F_8, etc. [51] doped with Er).

Unfortunately, the Stark structure of Yb^{3+} terms for these crystals has not yet been determined as of now. [59] is the only article to demonstrate splitting of the $^2F_{5/2}$ state of the Yb^{3+} ion (10,230, 10,367 and 10,587 cm^{-1}) for YF_3:Yb luminophors having a very sharply-defined structure of the excitation spectrum (in the 945 to 978 nm region).

Our IR spectrum research agreed satisfactorily with these data and permitted us to approximately estimate the Stark structure of terms for $NaYF_4$ bases, which were most effective for obtaining green anti-Stokes luminescence. The IR excitation spectrum for Y_2O_2S and La_2O_2S luminophors was shifted slightly towards the long-wavelength region (980 to 985 nm), and there is at 970 nm a wide gap (especially for Y_2O_2S and even more so as the temperature is lowered to 77 K), which degrades its spectral matching with GaAs LEDs. The lower Stark component of the $^2F_{5/2}$ excited state is located in these bases at $10,170 \pm 10$ cm^{-1} and the upper at $10,640 \pm 20$ cm^{-1}; overall splitting of the ground state is 480 ± 20 cm^{-1}. Finer splitting of the terms is made difficult by the overlapping of Er^{3+} ion absorption located in the same

37

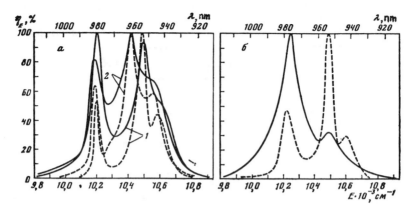

Fig. 16. IR Excitation Spectra for Visible Luminescence of Er^{3+} Ions in Oxysulfides (a: 1- Y_2O_2S, 2- La_2O_2S) and Fluorides (b: $NaYF_4$)

Continuous line is at 293 K, dashed line - 77 K

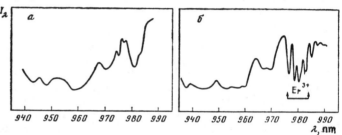

Fig. 17. Absorption Spectra of Oxysulfide Luminophors at 77 K

a- La_2O_2S:Yb,Er

b- Y_2O_2S:Yb,Er

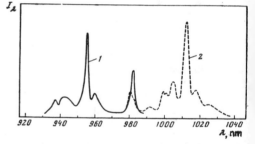

Fig. 18. Comparison of Excitation Spectrum for Anti-Stokes Luminescence (1) and IR Luminescence Spectrum (2) for Y_2O_2S:Yb,Er Luminophors

region of the spectrum, by the close spacing between lower components of the ground state ($\Delta E \approx 50$ cm^{-1}), and by the presence of centers with slightly different parameters. Besides, the accuracy we have achieved in determining the term structure is quite enough for a first-approximation estimate of the energy transfer probability between Yb^{3+} and Er^{3+} ions. For this, however, we need at least the same accuracy in determining the location of the terms of the Er^{3+} ions, which have a significantly richer complement of Stark components, requiring much colder temperatures for their determination. For YOCl bases the structure of the lower ($^4I_{15/2}$) and first excited ($^4I_{13/2}$) states were first determined in [59]. Our measurements [73,74] of the anti-Stokes emission spectra (Figs. 19,20), carried out at $T=293$, 77, and 4 K for YOCl:Yb,Er with laser pumping (in the range of $\lambda_B \approx 1.52$ μm), on the whole were in satisfactory agreement with these data.

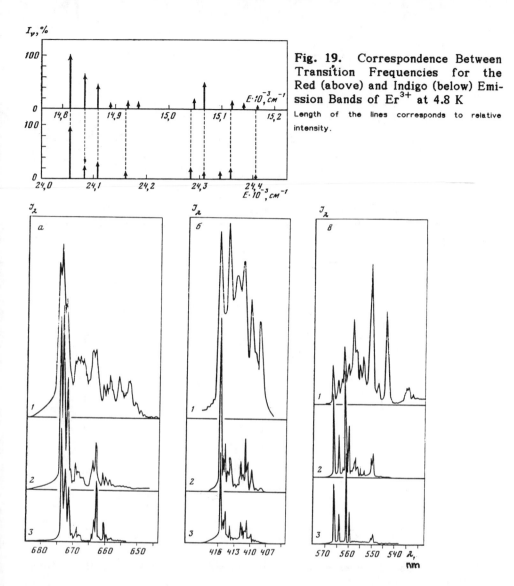

Fig. 19. Correspondence Between Transition Frequencies for the Red (above) and Indigo (below) Emission Bands of Er^{3+} at 4.8 K

Length of the lines corresponds to relative intensity.

Fig. 20. Fine Structure of the Anti-Stokes Emission Bands in YOCl:Yb,Er Luminophors With IR Pumping From a He-Ne Laser ($\lambda=1.15\ \mu$m) for Room (1), Nitrogen (2), and Helium (3) Temperatures. a- Red band ($^4F_{9/2}\rightarrow{}^4I_{15/2}$ transition), b- Indigo band ($^2H_{9/2}\rightarrow{}^4I_{15/2}$ transition), c- Green band ($^4S_{3/2}\rightarrow{}^4I_{15/2}$ and $^2H_{9/2}\rightarrow{}^4I_{13/2}$ transitions)

At $T=4.2$ K the structure observed for the indigo (406 to 415 nm) and red (650 to 670 nm) luminescence bands were practically identical, which allowed us to determine the Stark structure of the $^4I_{15/2}$ state of Er^{3+} (0, 30, 50, 105, 225, 248, 297 and 345 cm^{-1}). Overall splitting of this state was estimated to be 345 ± 10 cm^{-1} [4] instead of the 358 cm^{-1} in [59]. Our most recent reading of the IR absorption and IR luminescence spectra at 677 K

39

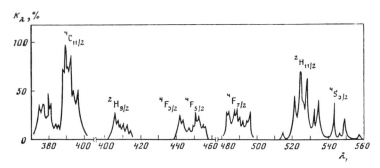

Fig. 21. Excitation Spectrum for Red Luminescence in YOCl:Yb,Er ($^4F_{9/2} \to$ $^4I_{15/2}$ transition) in the Stokes region. The terms produced by the excitation are indicated near the maxima.

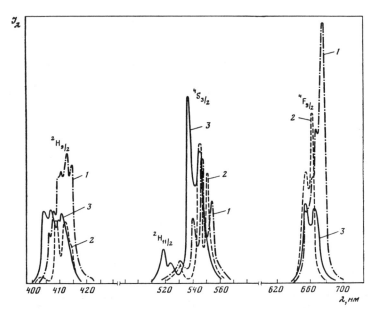

Fig. 22. Visible Luminescence Spectrum of Er^{3+} Ions in Various Bases During IR Pumping ($\lambda_B - 1$ μm)

1- YOCl; 2- Y_2O_2S; 3- $BaYF_5$. The terms associated with the transitions are indicated near the maxima.

yielded a value of $\Delta E = 365 \pm 10$ cm^{-1} [81]. More careful analysis of all our data allowed us to calculate $\Delta E = 354 \pm 5$ cm^{-1}; this is practically identical to the results of [59] and to the value for GdOCl $\Delta E = 355 \pm 10$ cm^{-1} obtained in [47]. Inhomogeneous line broadening (10 to 15 cm^{-1}) made it difficult to more accurately measure the location of levels, which could be displaced within those limits also by changes in activator or sensitizer concentration, or by the choice of method for synthesizing the luminophor.

Determination of the ground state structure of Er^{3+} allowed us to

---

[4]The location of the highest energy component in the emission spectra is less precise because of its strong reabsorption, especially at helium temperatures.

work from measurements of the absorption, luminescence and excitation spectra (Figs. 21,22,23), both Stokes and anti-Stokes, and find the location of the upper and lower Stark components of all states taking part in anti-Stokes luminescence with roughly the same accuracy (10 cm^{-1}). Our results for YOCl are organized in Table 4, which also presents all the data available from the literature on other oxychloride lattices (GdOCl and Y_3OCl_7). It is apparent from the table that the discrepancy with existing data on YOCl [59] is less than 10 cm^{-1}; it is in the range of 30 to 50 cm^{-1} for GdOCl [47], increasing to 50-150 cm^{-1} for Y_3OCl_7 bases, which can be completely explained by the substantial crystallographic differences between the YOCl and Y_3OCl_7 structures.

We conducted a similar but less detailed investigation of the fluoride bases $NaYF_4$ and BaY_2F_8 as well as oxysulfides of yttrium and lanthanum; the corresponding data for these bases are given in Tables 5 and 6, which also

Table 4. Energy Levels of Er^{3+} Ions (cm^{-1}) in Oxychloride Bases (Upper and Lower Stark Components)

Терм	YOCl	YOCl [59]	Y_3OCl_7 [59]	GdOCl [47]
$^4I_{15/2}$	0; 354	0; 358	0; 282	0; 335
$^4I_{13/2}$	6 560; 6 770	—	6 689; 6 804	6 497; 6 771
$^4I_{11/2}$	10 230; 10 332	10 238; 10 330	10 178; 10 286	10 159; 10 333
$^4I_{9/2}$	12 340; 12 585	—	12 430; 12 500	12 349; 12 572
$^4F_{9/2}$	15 172; 15 350	15 179; —	15 170; 15 662	15 173; 15 358
$^4S_{3/2}$	18 290; 18 396	—	18 266; 18 382	18 295; 18 408
$^2H_{11/2}$	18 950; 19 180	—	18 957; 19 135	18 945; 19 198
$^4F_{7/2}$	20 410; 20 640	—	20 346; 20 449	20 424; 20 625
$^4F_{5/2}$	22 010; 22 180	—	22 041; —	22 003; 22 167
$^4F_{3/2}$	22 350; 22 610	—	22 391; —	22 323; 22 589
$^2H_{9/2}$	24 414; 24 570	—	24 426; —	24 441; 24 586
$^4G_{11/2}$	26 150; 26 360	—	26 075; 26 302	26 182; 26 348

Table 5. Stark Components of Yb^{3+} Ion Terms (cm^{-1}) in Oxysulfide and Fluoride Bases

Term	Y_2O_2S	$NaYF_4$	YF_3	LaF_3	BaY_2F_8
$^2F_{7/2}$	0	0	0	0	0
	70	65	—	—	—
	190	210	—	—	—
	420	360	390	400	410
$^2F_{5/2}$	10 180	10 218	10 230	10 240	10 245
	10 510	10 440	10 367	—	—
	10 660	10 570	10 587	10 600	10 680

Table 6. Energy Levels of Er^{3+} Ions (cm^{-1}) in Oxysulfide and Fluoride Bases (Upper and Lower Stark Components of the Levels)

Терм	Y_2O_2S	NaYF₄	LiYF₄ [86]	YF₃ [59]	LaF₃ [51]	BaY₂F₈
$^4I_{15/2}$	0; 296	0; 290	0; 280	0; 423	0; 444	0; 410
$^4I_{13/2}$	6 480; 6 670	6 520; 6 730	6 511; 6 714	6 545; 6 766	6 604; 6 825	6 560; 6 900
$^4I_{11/2}$	10 210; 10 320	10 240; 10 350	10 222; 10 315	10 257; 10 369	10 302; 10 390	10 280; 11 390
$^4I_{9/2}$	—	12 370; 12 650	12 360; 12 660	—	12 420; 12 692	—
$^4F_{9/2}$	15 140; 15 310	15 320; 15 480	15 316; 15 477	15 339; 15 490	15 390; 15 522	15 310; 15 500
$^4S_{3/2}$	18 270; 18 360	18 450; 18 510	18 433; 18 494	18 494; 18 575	18 564; 18 594	18 490; 18 570
$^2H_{11/2}$	18 940; 19 160	19 170; 19 350	19 154; 19 338	19 200; —	19 270; 19 423	19 210; 19 400
$^4F_{7/2}$	20 400; 20 620	20 580; 20 690	20 565; 20 665	20 587; —	20 658; 20 791	20 590; 20 730

Dash indicates a lack of data

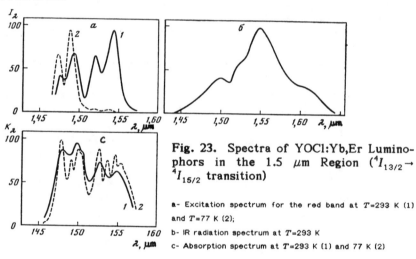

Fig. 23. Spectra of YOCl:Yb,Er Luminophors in the 1.5 μm Region ($^4I_{13/2} \to {}^4I_{15/2}$ transition)

a- Excitation spectrum for the red band at T=293 K (1) and T=77 K (2);

b- IR radiation spectrum at T=293 K

c- Absorption spectrum at T=293 K (1) and 77 K (2)

present data from the literature, including that for LiYF₄ laser crystals [86]. From the satisfactory coincidence of the aforementioned data on the closely-related crystallographic structures NaYF₄ and LiYF₄, we may determine (analogously to the case of YOCl and GdOCl) with some reliability and accuracy the size of the energy gap, which is required for estimating the various probabilities for energy transfer processes, and intra-center multi-phonon relaxation in various crystal lattices. This energy gap for multiphonon relaxation is changed much less than the inter-center transition energy.

The lower Stark component of the Er^{3+} ion excited state $^4I_{11/2}$ is situated in all the bases just a little higher than the lower Stark component of Yb^{3+} ion excited state $^2F_{5/2}$, which reduces the likelihood of Yb—Er energy transfer (and, consequently, efficient anti-Stokes luminescence at low temperatures).

However, this energy mismatch does not exceed 150 cm^{-1}, diminishing form 130 to 140 cm^{-1} for oxychloride bases to 30-50 cm^{-1} for fluoride bases. This is because at room temperature and higher, $\Delta E_0 < kT$ in all those bases. Therefore the condition of highly incoherent interactions of Yb^{3+} and Er^{3+} ions is satisfied in the one-photon state, which is corroborated by investigations of luminescence kinetics and the direct estimate which we will carry out in the following section. On the other hand, the mismatch in transition energies for the second stage of Yb—Er energy transfer and cross-relaxation is significantly higher and these processes require the emission (or absorption) of several lattice phonons. A direct correlation obtains here between the ratio of the energy differential (from the maximum energy an Er^{3+} ion can take on during transition from states $^4I_{13/2}$ to $^4F_{9/2}$ to the minimum energy a Yb^{3+} ion can transfer) to the phonon energy $N_\phi = \Delta E/\hbar\omega_\phi$, and the ratio of red and green band luminescence intensities.

Table 7. Effect of the Ratio of Transition Energy Mismatches During Summing in $^2F_{5/2} \to {}^2F_{7/2}$ (Yb) and $^4I_{13/2} \to {}^4F_{9/2}$ (Er); and During Cross-Relaxations $^4S_{3/2} \to {}^4I_{13/2}$ (Er) and $^2F_{7/2} \to {}^2F_{5/2}$ (Yb) to the Frequency of Lattice Phonons on the Color of Anti-Stokes Luminescence of Er^{3+} Ions

| Base | $\hbar\omega_\phi$ | Color of Emission | Summation | | | | Cross-Relaxation | | | |
			E_{min} delivered	E_{max} accepted	ΔE	$N_\phi = \dfrac{\Delta E}{\hbar\omega_\phi}$	E_{min} delivered	E_{max} accepted	ΔE	$N_\phi = \dfrac{\Delta E}{\hbar\omega_\phi}$
YOCl	620	Red	9380	8790	590	1	11 520	10 950	470	1
Y_3OCl_7	620	»	9520	8970	550	1	11 460	10 700	760	1
Y_2O_2S	520	Green-Red	9760	8870	920	2	11 600	10 660	940	2
LaF_3	350	Green	9820	8920	900	3	11 340	10 600	740	2
YF_3	375	»	9830	8940	890	3	11 630	10 590	1040	3
$NaYF_4$	350	»	9860	8950	910	3	11 720	10 570	1150	3—4

This correlation, first established in [68], is illustrated in Table 7 which shows that in lattices requiring emission of one, two or three phonons in order to excite the red emission band, that band will be, respectively, either dominant, roughly equal, or significantly weaker with respect to the green band. Due to thermalization in the excited state $^4F_{9/2}$ ($\Delta E \approx 200$ cm^{-1}) and the small population of the Yb^{3+} ground state upper Stark component at $T \leq 300$ K ($400 < E_0 < 700$ cm^{-1}) for cross-relaxation from this state, there obtains an even greater transfer energy mismatch (by 500 to 800 cm^{-1}). This requires an even greater quantity of phonons, necessitating not the emission but rather the absorption of two to four phonons, which is significantly less likely. Therefore the process of inter-ion cross-relaxation via this channel should not have a strong effect on the efficiency of red emission band in

43

Er^{3+} (the $^4F_{9/2}$ state may still decay with the emission of a phonon and transit to the ground state, but only with a very large energy deficit ($\Delta E > 4000$ cm^{-1}) and correspondingly low probability). The exact opposite happens for the green emission band. Actually, for that case, the second energy transfer from Yb^{3+} to Er^{3+} can lead to transition of the latter to any one of three states ($^4S_{3/2}$, $^2H_{11/2}$, or $^4F_{7/2}$), during which if transition to the first two states requires emission of several photons, then the $^4F_{7/2}$ state lies only a little higher (by 50 to 100 cm^{-1}) than the sum of the $^2F_{5/2}$ (Yb) and $^4I_{11/2}$ (Er) energies. Therefore the type of crystal lattice is not so critical for the given processes of cooperation of excitation energy at room temperatures. Meanwhile, for the dominant cross-relaxation processes (the $^4S_{3/2} \to {}^4I_{13/2}$ and $^2F_{7/2} \to {}^2F_{5/2}$ transitions), this energy mismatch is positive, in the range $500 \leq \Delta E \leq 1200$ cm^{-1} and it requires the emission of from one to four phonons, depending on the crystal lattice. Only the fluoride bases, where the green emission band predominated, required as a rule the emission of approximately three phonons, i. e. cross-relaxation processes are of lower probability. Only one phonon is enough in the same situation for oxychloride bases. In the given case there are some fine points to be noted, such as the fact that only two phonons are required for cross-relaxation in LaF_3 and thus that luminophor is weakly luminescent, while in $NaYF_4$ the ratio N_ϕ is close to 4 and thus it is an optimum luminophor for green emission. It is interesting that in Y_3OCl_7 the value of ΔE is somewhat higher than the phonon energy, and this may explain the fact that, for certain conditions of crystal growth and excitation, one can control the color of anti-Stokes luminescence [87]. Following this line of reasoning, we may also find the optimal concentration needed to obtain green emission with maximum efficiency (20 to 40% for $NaYF_4$; about 20% for YF_3; 12% for LaF_3 and only 8 to 10% for Y_2O_2S), which is limited by cross-relaxation processes.

3.2 **Kinetics of Afterglow in Cooperative Luminophors.** It is well-known [16,8,88] that the kinetics of afterglow in RE^{3+}-ion activated crystal phosphors can in no way be described for the general case by a simple exponential decay law. Instead of decay, an increase in afterglow is often observed after cessation of pulsed short wavelength excitation, which gives evidence of the inertial nature of intermediate energy transfer processes. Such an increasing afterglow may also arise during intra-center relaxation from intermediate higher-energy states. It is especially distinct during recombination and resonant energy transfer, which includes anti-Stokes pumping of cooperative luminophors. Generally speaking, the non-exponential nature of the decay processes may be caused by scatter in the values of the parameters of energy donor and acceptor pairs, including that arising from the difference in their inter-center distances [16,46]. However, thanks to a very high concentration of participating ions in cooperative luminophors, they are almost always located as close as allowable in the given crystal lattice and after some time their afterglow may become exponential. Nonetheless, to obtain reliable values for transition probabilities in this case too, it is necessary to draw up afterglow curves for a broad enough range of afterglow intensities, something which was not carried out in the first works on the subject of anti-Stokes luminescence kinetics.

The common characteristic of cooperative luminophors is that the one-photon excited state lifetime is substantially longer than the two-photon excited state lifetime. In addition, as was first noted by V. V. Ovsyankin and P. P. Feofilov [1,17], at low pumping intensities the luminescence decay after some time has elapsed may be described to a first approximation by an exponential with time constant $\tau_{AC} = \tau_0/M$, where τ_0 is the one-photon excited state lifetime and M is the number of summed excitations. This simple rule, proceeding from approximate time-dependent equations of balance, should be observed during cooperative sensitization. This is possible during sequential sensitization only if the lifetimes of the summed intermediate states are identical. In the most general case, a similar treatment gives $\dfrac{1}{\tau_{AC}} \sum_i \dfrac{1}{\tau_{i0}}$, where τ_{i0} is the one-photon excited state lifetime.

Table 8. Relaxation Times in Anti-Stokes Luminophors

Luminescence bands, transitions		YOCl [45]	Y₂O₂S [44]	NaYF₄ [85]	BaYF₃ [79]	LaF₃ [37]
IR Band						
Yb, $^2F_{5/2} \to {}^2F_{7/2}$		$4 \cdot 10^{-4}$	$3,5 \cdot 10^{-4}$	$2 \cdot 10^{-3}$	$2 \cdot 10^{-3}$	$2 \cdot 10^{-3}$
Er, $^4I_{11/2} \to {}^4I_{15/2}$		$2 \cdot 10^{-3}$	$2 \cdot 10^{-3}$	$1,2 \cdot 10^{-2}$	$1 \cdot 10^{-2}$	$1,2 \cdot 10^{-2}$
Yb and Er Bands at 1 μm	experimental	$4,4 \cdot 10^{-4}$	$3,9 \cdot 10^{-4}$	$2,9 \cdot 10^{-3}$	—	$2,2 \cdot 10^{-3}$
	calculated	$4,5 \cdot 10^{-4}$	—	$2,2 \cdot 10^{-3}$	—	$2,3 \cdot 10^{-3}$
Er at 1.5 μm		$5 \cdot 10^{-3}$	$2,9 \cdot 10^{-3}$	$1 \cdot 10^{-2}$	$1 \cdot 10^{-2}$	$1,2 \cdot 10^{-2}$
Green Band						
Intra-center for $C_{Er} \leq 1\%$		$1 \cdot 10^{-4}$	$1 \cdot 10^{-4}$	$4 \cdot 10^{-4}$	$5 \cdot 10^{-4}$	$1 \cdot 10^{-3}$
Cross-relaxation at C_{Er}, in %		3,3	6	3	100	—
		$6 \cdot 10^{-5}$	$5 \cdot 10^{-5}$	$1,8 \cdot 10^{-4}$	$4 \cdot 10^{-6}$	—
Cross-relaxation at C_{Er}, in %		10	10	20	100	12
		$2,5 \cdot 10^{-5}$	$2 \cdot 10^{-5}$	$7 \cdot 10^{-5}$	$1 \cdot 10^{-5}$	$1,3 \cdot 10^{-4}$
Relaxation during anti-Stokes pumping	experimental	$2,1 \cdot 10^{-4}$	$2,4 \cdot 10^{-4}$	$1,2 \cdot 10^{-3}$	—	—
	calculated	$2,1 \cdot 10^{-4}$	$2 \cdot 10^{-4}$	$1,4 \cdot 10^{-3}$	$9 \cdot 10^{-4}$	$1,1 \cdot 10^{-3}$
Red Band						
Intra-center at $C_{Er} < 1\%$	experimental	$3 \cdot 10^{-5}$	—	—	$1 \cdot 10^{-3}$	$1,9 \cdot 10^{-3}$ (экспериментальное)
	calculated					$7,5 \cdot 10^{-4}$ (расчетное)
Cross-relaxation at C_{Er}, in %		3	6	3	—	—
		$2,5 \cdot 10^{-5}$	$1 \cdot 10^{-4}$	$6 \cdot 10^{-4}$	—	—
Cross-relaxation at C_{Er}, in %		10	10	20	10	—
		$2 \cdot 10^{-5}$	$8 \cdot 10^{-5}$	$3,5 \cdot 10^{-4}$	$7 \cdot 10^{-4}$	—
Relaxation during anti-Stokes pumping	Initial stage	$2 \cdot 10^{-4}$	—	—	—	—
	Final Stage	$4,4 \cdot 10^{-4}$	$3,3 \cdot 10^{-4}$	$1,6 \cdot 10^{-4}$	$1,2 \cdot 10^{-3}$	—

Dash represents a lack of data; the concentration refers to the degree of trivalent lattice cation substitution.

The results of our research [45,68] and the data from the literature on excited state lifetimes and relaxation rates for cooperative luminophors with Yb and Er are presented in Table 8. A discussion of these data (generalized in part in the dissertation of A. A. Glushko [89]) starts with the simplest case of Stokes excitation of IR-band luminescence (Figs. 24,25).

What is most important is that conditions of incoherent interactions of sensitizer and activator ions in the one-photon excited state are usually met in cooperative luminophors. As was mentioned before, this leads to decay of both IR luminescence bands with the same time constant

$$\frac{1}{\tau} = \frac{P_{01}C_A\tau_A^{-1} + P_{10}C_S\tau_S^{-1}}{P_{01}C_A + P_{10}C_S}. \tag{63}$$

This property of IR-band afterglow during Stokes IR pumping in LaF_3:Yb,Er luminophors was first observed in [37], although for $BaYF_5$ the authors of [79] came to the conclusion that this equation was not satisfied. Meanwhile, our subsequent detailed investigation of YOCl:Yb,Er, as well as $NaYF_4$ and Y_2O_2S luminophors, by their agreement with the results of [44,85], confirmed this property for the majority of cooperative luminophors. This was also evidenced by numerical estimates which demonstrated that for $0.4 \cdot 10^{-4} \leq \tau_{Yb} \leq 2 \cdot 10^{-3}$s, $2 \cdot 10^{-3} \leq \tau_{Er} \leq 10^{-2}$s, $\lambda_B = 1.0$ μm, and $R_0 = 3.5$ Å, the product $P_{01}C_A\tau_S > 10$ and $P_{10}C_S\tau_A > 10$ for optimal concentrations $1 \leq C_A \leq 6\%$ and $8 \leq C_S \leq 40\%$.

In our paper for $C_{Er} = 3.3\%$ and $C_{Yb} = 10\%$, we obtained $\tau \approx 4.4 \cdot 10^{-4}$s, which is in satisfactory agreement with the estimate from the above formula. At concentration parity $C_{Yb} \approx C_{Er} \approx 3.3\%$ the average time is about $0.6 \cdot 10^{-3}$s, which allows us to estimate the ratio P_{01}/P_{10} for YOCl:

$$\frac{P_{01}}{P_{10}} = \frac{1 - \tau_A/\bar{\tau}}{\bar{\tau}/\tau_S - 1} \approx 1.4 \tag{64}$$

The most important qualitative confirmation that conditions of incoherent interactions are indeed met is the fact that the lifetime of energy donors (Yb^{3+}) does not decrease, on the contrary, it increases upon doping with acceptors (Er^{3+}), thereafter it would have to reduce, as in the case of weak interactions.

Since the lower excited state of Er^{3+} is in all $\Delta E = 140$ cm^{-1} lower than the lower excited state of Er^{3+}[sic], then in accordance with Eq. (40) at room temperatures ($kT = 210$ cm^{-1}) the value of ratio $1 < P_{01}/P_{10} < 2$, which is found to be in satisfactory agreement with experiment. The value of ΔE is even lower in oxysulfide and fluoride bases. In them, therefore, $P_{01}/P_{10} \approx 1$ at $T = 300$ K. Conversely, similar calculations for $NaYF_4$ luminophors show that $P_{01}/P_{10} < 1$. For even lower (helium) temperatures, obviously $\Delta E \gg kT$ and the energy transfer from sensitizer to activator (or vice-versa) may be "frozen". When this happens, conditions of incoherent interactions break down, which for YOCl luminophors is found to be in qualitative agreement with our experiments and which leads to substantial decrease in the luminescence efficiency.

Experimental data available at the present time attests that in all the bases the lifetime of Er^{3+} ion states responsible for visible luminescence is markedly shorter than the lifetime of Er^{3+} and even Yb^{3+} states responsible for energy summing processes. For the green luminescence band during Stokes pumping to the $^2H_{11/2}$ state, after a very brief build-up which occurs

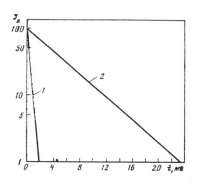 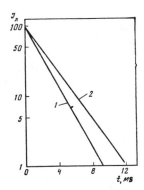

Fig. 24. Afterglow Kinetics From the $^2F_{5/2}$ Excited State of Yttrium (1), and the $^4I_{13/2}$ Excited State of Erbium (2) at $T=300$ K

Fig. 25. Afterglow Kinetics From the $^4I_{15/2}$ State of Erbium 1- T=300K; 2- T=77K

whether or not Yb^{3+} is present (the $^2H_{11/2}\rightarrow{}^4S_{3/2}$ relaxation), a simple exponential afterglow decay is observed (Fig. 26). According to our data for YOCl:Yb,Er the risetime does not exceed 5 μs, while the subsequent decay has a constant value of 60 μs at $T=300$ K (for $C_{Er}=3.3\%$ and $C_{Yb}=0$) and 25 μs at $C_{Yb}=10\%$. This data is in satisfactory agreement with the data of [59] where for $T=4.2$ K they obtained $\tau=100$ μs in the absence of Yb and $\tau=10$ μs for $C_{Yb}=10\%$. The reduction in the lifetimes of these states due to inter-ion cross-relaxation after Yb^{3+} doping was noted by practically all these authors since, as a rule, this lifetime is 2 to 3 times lower than it was initially in optimal-efficiency anti-Stokes luminophors. However, hardly any of these authors noticed that an increase in Er^{3+} concentration substantially shortened the lifetime of the given radiative state. This optimal Er^{3+} concentration also corresponds to a two-fold drop in τ, since overall τ decreases by a factor of 4 to 5 compared to τ for intra-center transitions.

It is noteworthy that this drop in τ is initiated at a lower concentrations of Er^{3+} than of Yb^{3+}. This demonstrates that cross-relaxation via two Er ions is more efficient than via Yb and Er ions. In particular, in the data of [79] for $BaYF_5$ luminophors, when there is almost total substitution of sensitizer ions for cations the lifetime of the $^4S_{3/2}$ falls off by a factor of 100, while for a similar substitution by activator ions it falls off by a factor af 200. This all takes place regardless of the fact that the transition probability in the Yb^{3+} ions is about 5 times higher than the probability of the $^4I_{11/2}\rightarrow{}^4I_{15/2}$ transition in Er, and hence this can only be explained as a change in the energy mismatch with a dominant channel for back transfer being the consequence. Actually there are three decay channels for the $^4S_{3/2}$ state (Er) with participation from a neighboring Er ion, all of which may have about the same probability:

a) $^4S_{3/2}\rightarrow{}^4I_{9/2};\quad {}^4I_{15/2}\rightarrow{}^4I_{13/2};$

b) $^4S_{3/2}\rightarrow{}^4I_{13/2};\quad {}^4I_{15/2}\rightarrow{}^4I_{11/2};$

c) $^4S_{3/2}\rightarrow{}^4I_{11/2};\quad {}^4I_{15/2}\rightarrow{}^4I_{13/2}.$

47

Of these channels, only b) can be realized with the participation of Yb^{3+} ions, since the others would require absorption of 5 to 10 phonons, which is highly unlikely. This confirms anew that the role of the sensitizer is basically to increase the ratio of probabilities of forward and back transitions, which in the meantime also still depend on the crystal lattice.

The kinetics of Er^{3+} green-band luminescence during anti-Stokes pumping is usually well described by the equation

$$J = \frac{P_S N_{S0} P_{01} P_{10} C_A C_S}{(P_{01} C_A + P_{10} C_S)^2 (a - 2\tau^{-1})} (e^{-2t/\tau} - e^{at}),$$

(65)

where P_S is the coefficient of pumping energy cooperation with participation of the sensitizer, P_{01} and P_{10} are the coefficients of energy transfer from sensitizer to activator and back, C_A and C_S are their concentrations, N_{S0} is the initial concentration of excited sensitizer ions, and $a = \frac{1}{T} + Q_S C_S + Q_A C_A$ is the total decay probability of the radiative state $^4S_{3/2}$ as a result of intra-center and inter-ion cross-relaxation processes.

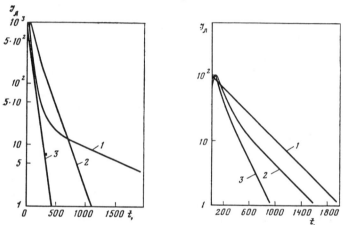

Fig. 26. Kinetics of Afterglow From the $^4F_{9/2}$ (1, 2) and $^4S_{3/2}$ (3) States of Erbium Ions   1, 3- T=300 K, 2- T=77 K

Fig. 27.   Kinetics of Afterglow From the Excited States of Yttrium and Erbium to the Doubly Activated System $Y_{0.87}Yb_{0.12}Er_{0.01}OCl$ 1- $^2F_{5/2}$ (Yb^{3+}) and $^4I_{11/2}$ (Er^{3+}) states; 2- $^4F_{9/2}$ state (Er^{3+}); 3- $^4S_{3/2}$ state (Er^{3+})

In this way the afterglow build-up of the green band during anti-Stokes pumping is determined by the total decay probability of this state, which may be measured independently of the experiments with Stokes pumping, and the subsequent decay corresponds to the intermediate state half-life, which may be found from the decay of IR band luminescence. Such kinetic behavior is observed in experiments for YOCl and other bases, which confirms the universal nature of the excitation and relaxation mechanism.

The afterglow kinetics for the Er^{3+} red band (the $^4F_{9/2} \rightarrow {}^4I_{15/2}$ transition) under anti-Stokes and even under Stokes pumping are significantly more complicated (Figs. 26,27). As pointed out earlier, there have been six different schemes proposed by various authors for interpreting the kinetics of the intermediate energy transfer processes. For YOCl bases, which we researched in some detail, red band decay is preceded by a relatively long build-up, extending from 25 to 30 μs for Stokes pumping to 80-85 μs for anti-Stokes pumping. During Stokes pumping (luminophors without Yb) the experimental decay curve has a later exponential section with $\tau \approx 60$ μs, but then the decay is moderated substantially and at later stages has $\tau \approx 1.2$ μs. In luminophors with Yb those time constants take on values of, respectively, 30 and 200 μs. For anti-Stokes pumping of these luminophors the smaller time constant is about 200 μs and the time constant for later stages is equal to 420 μs. Trying to understand such a complicated picture without prior knowledge about the possible mechanisms for populating the given radiative states would be very difficult.

Using the experimental data on the afterglow of other bands, it can be relatively easy to ascertain which of the possible mechanisms is actually at work in a given system. We will start with the simpler asymptotic behavior at the later stages of decay. During anti-Stokes pumping this behavior directly corresponds to the simple summing of two excited states ($^2F_{5/2}$ for the Yb^{3+} ion and $^4I_{13/2}$ for the Er^{3+} ion) with substantially differing lifetimes. In fact, $\tau_{AC}^{-1} = \tau^{-1}(^4I_{15/2}) + \overline{\tau}^{-1} \approx (420 \ \mu s)^{-1}$, and this is in close conformity with experiment, which confirms this model of $^4F_{9/2}$ state population as a result of the summing of the aforementioned intermediate states of Yb and Er. It is interesting to note that, according to our data, summing of one-photon excited states of Er occurs also during Stokes pumping, after which a backlog of high-energy particles is exhausted as a result of inter-ion cross-relaxation. The value $\tau = 1.2$ ms here is in good agreement with $\tau^{-1} = \tau^{-1}(^4I_{13/2}) + \tau^{-1}(^4I_{11/2}) = (1.4 \ ms)^{-1}$, and a similar mechanism for the green band is lacking, which indicates dominance of relaxation processes with formation of a $^4I_{13/2}$ state.

At earlier stages of anti-Stokes-pumped afterglow the decay kinetics of the red and green bands are a little closer to each other, attesting to the benefits of populating the $^4F_{9/2}$ state by means of multiphonon cross-relaxation from the $^4S_{3/2}$ state. The mechanism of summing three excitations may be excluded for the given lattice, insofar as the red emission quantum yield depends in a linear way on I. In this manner, for short pumping durations a switching of the primary channels for populating the $^4F_{9/2}$ state in the decay process occurs, which is easy to understand considering the inherent delay time for populating the $^4I_{13/2}$ state (>0.5 ms).

This nontrivial case is realized at initial stages of decay during what would seem to be the simplest Stokes pumping of the red band. Actually we obtained a surprisingly close coincidence in delay times for both bands, in luminophors with and without Yb, whereas per the discussion on the term energy structure, doping should strongly quench the green band. This contradiction is easily removed if we consider that $\tau(^4F_{9/2}) \ll \tau(^4S_{3/2})$ and that during Stokes pumping the $^4F_{9/2}$ state is initially populated through $^4S_{3/2}$. Then a simple examination will show that, in any case, both bands should decay with the same time constant $\tau(^4S_{3/2})$. We can find the value of $\tau(^4F_{9/2})$ itself from the position of the maximum of the build-up curve, t_{max}. For a numerical calculation the corresponding formula may be conveniently described in the form

$$\frac{t_{max}}{\tau(^4S_{3/2})} = \frac{\ln(1+x)}{x} \quad , \tag{66}$$

where

$$x = \frac{\tau(^4S_{3/2})}{\tau(^4F_{9/2})} - 1.$$

Substituting into this $\tau(^4S_{3/2}) = 60$ μs and $t_{max} = 25$ to 30 μs, we get $\tau(^4F_{9/2}) = (15\pm3)$ μs, which is entirely possible considering both the higher intra-center multiphonon relaxation of this state, and inter-center cross-relaxation.

Investigation of afterglow kinetics also permits understanding of the primary mechanism populating the $^4F_{9/2}$ state during pumping in the 1.5 μm range, which corresponds to absorption of IR radiation by the radiating Er ions themselves. If it is populated as a result of summing the energy of three singly-pumped ions (in the $^4I_{13/2}$ state) and subsequent relaxation from the $^4S_{3/2}$ state, then the decay will have a time constant equal to $\frac{1}{3}\tau(^4S_{3/2})$, i.e., 1.6 ms for YOCl. However, our experiments showed this constant to be around 0.5 ms, i.e. somewhat lower than $\bar{\tau}(Yb,Er)$ for luminophors with identical concentrations of Yb and Er (0.6 ms for $C_{Yb} = C_{Er} = 0.1$). From this most simple treatment it follows that for proposed mechanism of populating the given state with the participation of Yb^{3+} ions, the afterglow delay time should be determined as

$$\frac{1}{\tau} = \frac{1}{\bar{\tau}_{(Yb,Er)}} + \frac{1}{\tau(^4I_{13/2})} \approx 0.52 \text{ ms} \tag{67}$$

which is found to agree closely with experiment.

At the maximum pumping densities which we used (10 W/cm²) this characteristic time even decreased (for initial afterglow stages up to 0.4 ms), which is naturally explained by the reduction in the $^4I_{13/2}$ state lifetime due to pumping energy cooperation processes. A similar reduction of the afterglow delay time with simultaneous decrease of the exponent in the dependence of J on I was observed by us during excitation by the focused beam of a YAG:Nd^{3+} laser ($I \approx 10^4$ W/cm²).

This is why it is necessary to consider the pumping intensity when measuring excited state lifetimes, and at very high pumping intensities we should expect, instead of exponential, a hyperbolic decay law similar to that in recombination luminescence. The presence of quenching impurities leads to a curtailing of intermediate state lifetimes. In particular, as our experiments demonstrated, doping Dy^{3+} ions into YOCl:Yb,Er luminophors in 0.1% concentrations, which led to an order of magnitude quenching of the anti-Stokes luminescence brightness, also resulted in an almost two-fold decrease in the afterglow duration.

We will look into the question of the relationship of probabilities for intra-center multiphonon radiative transitions in a following section with a discussion of experimental results on radiative efficiency during Stokes and anti-Stokes pumping.

3.3 **Energy Characteristics of Cooperative Luminophors.** As has been noted more than once, in the literature one meets with widely disparate data as the result of experiments [1-3,49,90] and theoretical estimates [12,21,31,91-93] on the absolute efficiency of anti-Stokes luminescence in RE^{3+} ions. In this section we will present the results of our measurements of the quantum efficiency η^* and the quantum yield η for the given class of luminophors during Stokes and anti-Stokes pumping.

Measurements were for the most part conducted using an integrating sphere according to the method described in Section 1. GaAs:Si LEDs, developed at GIRedmet [76], were employed as sources of IR excitation; later on we used commercially-available LEDs (AL-107B) as well as incandescent lamps with a selection of glass and interference light filters. Maximum IR pumping density was (0.9 ± 0.1) W/cm^2 with the incandescent lamp and (0.6 ± 0.1) W/cm^2 with the LEDs. In the latter case, the IR-to-visible conversion efficiency was found to be in agreement with the results of independent, although less accurate, measurements of the absolute luminescence brightness of such LEDs (coated with anti-Stokes luminophors) at the given intensity of their IR radiation.

Tables 9 and 10 present the efficiencies of anti-Stokes luminescence bands for various cooperative luminophors during pumping by an incandescent lamp, since in that case the IR pumping density is determined more accurately, and the luminophors operate in a more suitable and easily-standardized regime ("reflection") for an infinitely thick layer.[5]

It is apparent from Tables 9 and 10 that the greatest efficiency $\eta^* = (1.4 \pm 0.2)\%$ for red emission is achieved in oxychloride bases. It drops off by approximately an order of magnitude in oxysulfide bases, decreasing by that much again in fluoride bases. The absolute efficiency of green luminescence is noticeably less dependent on the crystal lattice and varies by no more than a factor of 10, so the maximal value $\eta^* = 0.2\%$ reached in oxysulfide (Y_2O_2S) and fluoride ($NaYF_4$) bases is roughly 7 times lower than that for the red emission band in oxychloride bases. The indigo band, which requires the summing of at least three excitations for its generation, is weaker yet. Its efficiency varies by little more than an order of magnitude in all the investigated bases, $10^{-3} \leq \eta \leq 10^{-2}\%$.

The results presented herein are in satisfactory agreement with some of the data encountered in the foreign literature where, for example, it is indicated that the quantum efficiency of $(Y_{0.79}Yb_{0.2}Er_{0.01})F_3$ luminophor pumped by a GaAs diode with power density 1 W/cm^2 is 0.04% [2]. The intensity ratios of the green and red band in various bases are also in satisfactory agreement with data in the literature [2,3]. This attests to the fact that domestic cooperative luminophors can hold their own with the best of that kind from abroad.

We will remark on two circumstances which in the final analysis are connected with the differences in the excitation spectra of various crystal bases. The first is that an oxychloride base has a much broader excitation spectrum and, consequently, better and more consistent agreement with the excitation spectrum of the source. The second is that, in developing new luminophors, the authors inadvertently tended to provide optimal pumping conditions for the new luminophors, not for the earlier-synthesized "standard" samples. Thus, for example, an increase in the efficiency of a green luminophor at the excitation spectra maxima is usually accompanied

---

[5]Using an interference reflective coating for thin screens ($h\approx20$ mg/cm) allowed us to obtain, in light of the nonlinearity of cooperative luminophors, a much higher (by a factor of 1.5 to 1.7) efficiency compared to the "infinitely" thick packing; however, it made a quantitative estimate of absorbed power more difficult.

Table 9. Quantum Efficiencies of Anti-Stokes Luminescence (in %) for Er^{3+} and Yb^{3+} Ion Pairs in Various Bases at $\lambda_B = 0.98 \pm 0.06$ μm ($T = 295$ K, $I = 1$ W/cm^2)

Base	Emission Band			Base	Emission Band		
	Red	Green	Indigo		Red	Green	Indigo
YOCl	1.4	0.03	$6 \cdot 10^{-3}$	NaYF$_4$	0.012	0.17	$1 \cdot 10^{-3}$
YOF	0.9	0.02	$5 \cdot 10^{-3}$	BaYF$_5$	0.01	0.06	$7 \cdot 10^{-4}$
La$_2$O$_2$S	0.12	0.11	$2 \cdot 10^{-3}$	YF$_3$	0.01	0.03	$5 \cdot 10^{-4}$
Y$_2$O$_2$S	0.11	0.19	$3 \cdot 10^{-3}$				

Table 10. Effect of Temperature on Quantum Efficiency of Anti-Stokes Luminescence of YOCl:Yb (11%), Er (9%) Luminophor at $\lambda_B = 1.52$ μm, $I = 10$ W/cm^2

Luminescence Band	Efficiency (in %) at Temperature, K		
	295	77	4.2
Red	0.5	2.5	0.35
Green	$2 \cdot 10^{-3}$	0.02	$1.4 \cdot 10^{-3}$
Indigo	$5 \cdot 10^{-4}$	$6.5 \cdot 10^{-3}$	$6 \cdot 10^{-4}$

by a narrowing of the spectrum itself, i.e. an increase is attained in the maximum value of the coefficient of absorption and, consequently, in the volume pumping density. As a result, NaYF$_4$ and Y$_2$O$_2$S luminophors have at most 4 to 5 times the gain compared to La$_2$O$_2$S when selectively pumped, although when a more wideband source is used this becomes no more than a factor of 2. At the same time, direct measurements of the quantum yield are to a significant degree complicated by the partial overlap of the IR-excitation and IR Stokes luminescence spectra of Yb^{3+} and Er^{3+} ions, which may lead to some incorrect conclusions about a weaker absorption of pumping radiation than is actually the case. It is therefore not surprising that, in just multiplying the efficiency ratios taken from promotional announcements for various new luminophors, one can get the idea that subsequent development has increased efficiency by 100 times or so. Our direct comparisons of samples from 1970 to 1980 showed however that over this time the value of η^* increased by a factor of only 3 to 4. For the composition of these luminophors most of the improvement is due to increased purity of the initial material (going from "4 nines" (99.99%) to "6 nines" (99.9999%)) and the corresponding reduction of effective quenchers such as Dy^{3+} and Fe^{2+}[90].

52

Meanwhile, it is too much to expect that further increases in materials purity beyond what is presently available ("6 nines") will lead to significant gain in anti-Stokes luminescence efficiency. This conclusion is supported by our direct measurements of the quantum yield of Yb and Er IR luminescence ($\eta_{ик}$) during Stokes pumping (GaAs laser at $\lambda = 0.91$ μm). In YOCl bases this value is 0.84 ± 0.07 for Yb^{3+} concentrations in the range 3.3 to 22%, dropping to $\eta_{ик} = 0.6$ at $C_{Yb} = 33\%$.

Close values for $\eta_{ик}$ (0.7 to 0.9) were obtained also for other bases (Y_2O_2S and $NaYF_4$) at optimal concentrations, from the standpoint of anti-Stokes luminescence efficiency, of activator and sensitizer ions. The scatter of values was not for the most part determined by experimental error (no more than 3-5% for the measured quantity), but rather the variations in samples of nominally identical composition, but prepared from various batches of raw materials not always synthesized under identical conditions (temperature and duration of firing, volume of the batch). Apparently this scatter is caused by differing concentrations of quenching centers (residual impurities, micro- and macroscopic lattice defects), to which excitation energy may migrate if the concentration of participating ions is high enough.

Nonetheless, on the basis of these experiments we may consider that the radiative lifetime τ_{Yb} is no more than 15% higher than experiment, and so the indicated quenching plays no substantial role. Thus further improvement in the technology of preparing these luminophors might improve their efficiency by only a few dozen percent. This is also directly confirmed by experiments measuring the summed quantum yield $\Sigma \eta_i$ of visible and IR-band luminescence during Stokes pumping of Er^{3+} ions to the $^4S_{3/2}$ (or $^4H_{11/2}$) state and the $^4F_{9/2}$ state, which are responsible for, respectively, the green or red luminescence bands during IR anti-Stokes pumping. The value of $\Sigma \eta_i$ for Y_2O_2S and La_2O_2S luminophors lies in the range 1.4-1.6, i.e. markedly greater than unity. This is why the summed radiative transition efficiency in the IR and visible regions of the spectrum are fairly great in these luminophors. Along with this, as a result of inter-ion cross-relaxation and intra-center multiphonon relaxation, the visible band efficiency is markedly lower than one would expect from the optimized model of Section 2.

We determined the pumping energy absorption for several of the most efficient cooperative luminophors. It turned out that, for YOCl, Y_2O_2S and $NaYF_4$ luminophors of optimal composition, at absorption spectrum maxima the diffuse reflection of pumping radiation was only 10-15%. Consequently, it may be that for selective pumping the quantum luminescence efficiency is close to the quantum luminescence yield. Unfortunately, however, there are no pumping sources available at the present time that are selective enough as well as powerful, aand that have a spectrum which coincides with the principal excitation maxima of cooperative luminophors. Approximate calculations showed that, for the IR lamp excitation that we used, the quantum yield of anti-Stokes luminescence is 2.5 to 3 times higher in YOCl luminophors, 3.5 to 4 times higher in Y_2O_2S, and even 5 to 6 times higher in $NaYF_4$, than the quantum luminescence efficiencies presented in Tables 9 and 10. This is why the quantum yields of the best cooperative luminophors lie in the range $1 < \eta < 10\%$ for $I = 1$ W/cm^2, i.e. almost the same range as for the maximum values given in the approximate theoretical estimate (cf. Section 2).

The subject of the quantum yield, important in both scientific and practical regards, will be examined in more detail by considering intra-center multiphonon relaxation processes and the existence of multiple channels for summing and inter-ion cross-relaxation for the green and red radiative bands of Er^{3+} ions. We will carry out parallel estimates for two luminophors — the YOCl luminophor, studied in more detail in this article, for which the red anti-Stokes emission quantum yield is higher than all the other luminophors (up to 2-3% at $I=1$ W/cm^2), and the significantly less efficient LaF$_3$:Yb,Er luminophor ($\eta \approx 0.1\%$ at $I=1$ W/cm^2) with predominantly green luminescence. The choice of the latter luminophor was fixed by the fact that almost all the parameters needed for estimating the anti-Stokes luminescence yield are known, thanks to the work of M. J. Weber [55,56] and other authors [57,58]. Besides this, carrying out estimates for cooperative luminophors which are so markedly different is of interest as regards the capability of current theory to explain these differences.

Above all for such estimates, in addition to the energy gaps between the various states, we need the values of the constants α and β which determine, respectively, the probability of intra-center multiphonon relaxation and the probability of lattice phonon-assisted energy transfer processes. According to the experimental data of various authors [12], for LaF$_3$ the value of $\alpha = 5 \cdot 10^{-3}$ cm which, in light of the relationship adduced earlier $\beta = \alpha - (\hbar\omega_\Phi)^{-1}\ln 2$ (where $\hbar\omega_\Phi$ is given in cm^{-1}) should yield $\beta = 3 \cdot 10^{-3}$. In articles [27,29] from the results of a calculation on temperature dependence of YF$_3$, with a lattice similar to that of LaF$_3$, a somewhat higher value ($\beta = 4 \cdot 10^{-3}$) was presented. In our calculations for LaF$_3$ luminophors we used the values $\alpha = 5 \cdot 10^{-3}$ cm, $\beta = 3.5 \cdot 10^{-3}$ cm and $w_\Phi(0) = 10^8$ s^{-1}. According to all the available data [26,28], the values of α and β are lower for YOCl luminophors than for fluoride bases (where they are higher than in any other luminophor), however they are determined less accurately and lie in the rather broad ranges $2 \cdot 10^{-3} \leq \alpha \leq 4 \cdot 10^{-3}$ cm and $1 \cdot 10^{-3} \leq \beta \leq 2 \cdot 10^{-3}$ cm. The best agreement with our experimental data on the yield of Stokes-pumped YOCl emission bands was for $\alpha = 4 \cdot 10^{-3}$ cm and $w(0) = 10^8$ s^{-1}, which by virtue of the connection between α and β gives $\beta = 2.5 \cdot 10^{-3}$ cm. Using a higher value for $w(0)$ or a lower α, as was done by a number of authors [31,59], leads to contradictions with experimentally measured lifetimes of the $^4I_{11/2}$ state (2 ms at $T = 293$ K and 2.6 ms at $T = 77$ K), the ratio of IR transitions in Er^{3+} ions ($J(1.1 \ \mu m)/J(1.5 \ \mu m) \approx 1$ at $T = 293$ K), and the lifetime of the $^2H_{9/2}$ state. In this way, the expected quantum yields of the red and green luminescence bands of Stokes-pumped YOCl bases is, neglecting inter-ion cross-relaxation, 0.85 and 0.35 respectively, which is found to be in good agreement with our experiments (0.7-0.8 for the green and 0.25-0.3 for the red band at low Er concentrations). In view of the high α in LaF$_3$ luminophors, these values should be higher yet, being, in accordance with [55,56], about 0.9 for transitions from the $^4S_{3/2}$ state (green band) and about 0.5 for transitions from the $^4F_{9/2}$ state (red band). In this way, multiphonon intra-center relaxation hardly affects the absolute efficiency of the green emission band, while noticeably lowering it for the red band (but no more than a factor of 2 to 3, which differs sharply from the conclusions in [31] where the multiphonon relaxation probability, instead of being compared to actual lifetimes for these states, was compared to an averaged value which was taken to be 10^{-1} for all transitions).

We discuss now the fundamental results from estimates of coefficients of pumping energy cooperation and inter-ion cross-relaxation in these luminophors. During this we will take the value of $P_S(0)$ in YOCl equal in the

Table 11. Probability of Multiphonon Relaxation w_Φ and Total Transition Probability w_Π for Er^{3+} Ions in YOCl

Initial Term	Next lower Term	ΔE (experimt'l) cm^{-1}	w_Φ (calculated) s^{-1}	w_Π (experimt'l) s^{-1}	$\dfrac{\Delta w}{w_\Pi}$	η
$^4I_{11/2}$	$^4I_{11/2}$	3460	$2\cdot10^2$	$5\cdot10^2$	0,6	0,5
$^4S_{3/2}$	$^4F_{9/2}$	2920	$1,5\cdot10^3$	$1\cdot10^4$	0,85	0,7
$^4F_{9/2}$	$^4I_{9/2}$	2580	$7\cdot10^3$	$1,5\cdot10^5$	0,5	0,4
$^2H_{9/2}$	$^4F_{3/2}$	1800	$1\cdot10^5$	$1\cdot10^5$	0,1	0,1

The calculation was carried out according to the formula $w_\Phi = w(0)(n+1)^N \exp(-\alpha\Delta E)$, and: $w(0) = 10^8$ s^{-1};

$$n = \left(e^{\frac{\hbar\omega_\Phi}{kT}} - 1\right)^{-1}, \quad \hbar\omega_\Phi = 500 \text{ cm}^{-1}; \quad kT = 210 \text{ cm}^{-1}; \quad N = \frac{\Delta E}{\hbar\omega_\Phi}; \quad \alpha = 4\cdot10^{-4} \text{ cm}.$$

Table 12. Coefficients of Cooperation of Pumping Energy (P) and Inter-Ion Cross-Relaxation (Q) for YOCl:Yb,Er and LaF$_3$:Yb,Er at $T = 300$ K

Base	Anti-Stokes Luminescence	Presence of Sensitizer	Cooperation ΣP_i, cm^3 s^{-1}	Cross-relaxt'n ΣQ_i, cm^3 s^{-1}	$\dfrac{\Sigma P_i}{\Sigma Q_i}$
YOCl	Green band $^4S_{3/2} \rightarrow {}^4I_{15/2}$	w/o Yb	$1\cdot10^{-16}$	$2\cdot10^{-17}$	5
		w/ Yb	$1\cdot10^{-15}$	$2\cdot10^{-16}$	5
	Red band $^4F_{9/2} \rightarrow {}^4I_{15/2}$	w/o Yb	$1\cdot10^{-17}$	$3\cdot10^{-18}$	3
		w/ Yb	$3\cdot10^{-16}$	$2\cdot10^{-19}$	$1.5\cdot10^3$
LaF$_3$	Green band $^4S_{3/2} \rightarrow {}^4I_{15/2}$	w/o Yb	$1\cdot10^{-18}$	$2\cdot10^{-18}$	0.5
		w/ Yb	$2\cdot10^{-17}$	$1\cdot10^{-18}$	20
	Red band $^4F_{9/2} \rightarrow {}^4I_{15/2}$	w/o Yb	$1.5\cdot10^{-19}$	$5\cdot10^{-20}$	3
		w/ Yb	$5\cdot10^{-19}$	$5\cdot10^{-20}$	10

first approximation to 10^{-15}cm$^3\cdot$s^{-1} (i.e., the same order of magnitude as Kushida's theoretical calculation and our estimate based on the classical Förster-Dexter-Galanin formula of $\tau_A = \tau_S = 10^{-3}$s), in accordance with the average values as presented by Mita [26]. These estimates are carried out for LaF$_3$, taking into account actual lifetimes τ in a ratio with various transitions and the wavelengths corresponding to those transitions, according to the formula

$$P_S(0) = 10^{-15} \frac{\bar{\lambda}_A^3 \bar{\lambda}_S^3}{\tau_A \tau_S} \text{ (cm}^3\cdot\text{c}^{-1}), \tag{68}$$

where τ is in milliseconds, and λ in microns. The results of calculation (Tables 11 and 12) qualitatively and in part quantitatively agree with the experimental data on absolute efficiency of the red and green emission bands, as well as the reduction in the $^4I_{9/2}$ and $^4S_{3/2}$ state lifetimes in the

55

investigated cooperative luminophors. Actually, the highest ratio of $P_S/Q_S \approx 1.5 \cdot 10^3$ (at the same time as $P_A/Q_A \approx 5$) was realized for the red emission band in YOCl, with which limiting values of D must also be determined at high concentrations of particpating ions. Considering $\sigma\tau = 1 \cdot 10^{-23}$, $I = 5 \cdot 10^{18}$ photons$\cdot$cm$^{-2} \cdot$s^{-1} (i.e. 1 W/cm^2; $\lambda_B = 1$ μm), this gives a maximum value for $\eta = 7.5\%$, or taking multiphonon relaxation into account $\eta = 2$ to 3%, which is found to be in close accord with experiment. For the green emission band in this same luminophor, the elementary approximations $P_S/Q_S \approx P_A/Q_A = 5$, i.e. in accordance with experiment, a substantially lower value of D and η and decrease in $\tau(^4S_{3/2})$ upon doping with sensitizers. The calculated $\eta = 0.02\%$ is somewhat lower than experiment (0.03%), which attests to the benefit of a lower value for Q_S (in particular, kinetics experiments give $Q_S \approx Q_A \approx 10^{-18}cm^3 \cdot$c^{-1}).

For the green emission band in LaF$_3$ we obtained a somewhat more favorable ratio of $P_S/Q_S = 20$ (for $P_A/Q_A \leq 1$) and $\eta = 0.1\%$, which was in satisfactory agreement with experimental values for efficiency. The absolute values P_S and $P_A \approx 1.5 \cdot 10^{-18}cm^3 \cdot$s^{-1} also were in satisfactory agreement with experimental data on the reduction in $\tau(^4S_{3/2})$ during activator and sensitizer doping. For the red emission band the ratio $P_S/Q_S \gg P_A/Q_A \approx 1$ and even reaches $1 \cdot 10^4$, i.e. larger than for YOCl. However, the absolute value itself of $P_S \approx 1 \cdot 10^{-18}cm^3 \cdot$s^{-1}, i.e. almost three orders of magnitude lower than YOCl, and thus the efficiency of red emission in this luminophor is primarily limited by the small value of P_s, not cross-relaxation processes, which is caused by a relatively great mismatch in transition energies. The above approximations were confirmed by direct numerical calculations of the equations of detailed balancing on a computer, allowing us simultaneously to consider both anti-Stokes luminescence channels in each luminophor and to vary the transition constants to get closer agreement with the experimental data. As a result of these calculations a complete set of transition probabilities were refined in accuracy, giving better agreement with experimental data on η for YOCl:Yb,Er luminophors (Tables 13,14).

We will not dwell in detail on the results of our measurements of the dependence of luminescence efficiency η^* on the concentration of participating ions (C_A and C_S), puming intensity I and temperature T, inasmuch as these results were on the whole in satisfactory agreement with the data in the literature. In accordance with the model we investigated, the maximum degree of nonlinearity K_H in the functions $J(C_S)$ and $J(I_{\text{ИК}})$ was no higher than the number of summed one-photon excitations. However, the limiting value of K_H was realized only at extremes of sensitizer concentration and pumping intensity. The transition to a linear dependence on I is fairly gradual, as is graphically illustrated in Fig. 28, which shows the dependence of the degree of nonlinearity on pumping intensity for various anti-Stokes emission bands pumped in the 1.5 μm range. Although the green emission band by energy considerations corresponds in this case to summing of three excitations, the fact that its degree of nonlinearity is a little more than 3 does not contradict what we said earlier about the limiting nonlinearity of a band. This is easily explained by the fact that, as has been already noted, in addition to the fundamental emission band (the $^4S_{3/2} \rightarrow {}^4I_{15/2}$ transition) in the green part of the spectrum, a second band is observed (the $^2H_{9/2} \rightarrow {}^4I_{13/2}$ transition) which, like the blue band (the $^2H_{9/2} \rightarrow {}^4I_{15/2}$ transition) corresponds to the summing of four excitations. As was remarked on earlier (cf. Section 2), in η as a function of concentration maxima are obtained both with respect

56

to C_S and with respect to C_A. In addition to depending on inter-ion cross-relaxation probabilities and one-photon excitation transfer, the location of these maxima also depend on the initial purity (Fig. 29) and the wavelength of pumping IR radiation even at the extremes of the same absorption band. Thus, for example, the optimal C_S values are 18, 22 and 27% for YOCl:Yb,Er luminophors pumped by, respectively, an incandescent lamp in the maximum Yb^{3+} absorption band, a GaAs laser (0.91 μm) and a YAG:Nd laser (1.06 μm). These differences may be qualitatively explained by considering the different absorption coefficients for each of the aforementioned cases and the reabsorption of visible anti-Stokes luminescence. The temperature dependence of anti-Stokes luminescence efficiency is significantly less well-defined than in temperature-sensitive crystal phosphors with recombination radiation mechanisms [41]. According to the experimental data of many authors [2,3,59], at higher than room temperatures there is usually observed a slow monotonic increase in luminescence efficiency, but for cooling to 80-150 K a moderate rise (a few times larger) changing to an even sharper drop-off for further decrease in temperature (down to 4.2 K). I. I. Sergeyev in [31] managed to obtain such behavior for $\eta(T)$ by considering the dependence of the

Fundamental Parameters of Participating Ions in YOCl:Yb,Er Luminophors Determining the Efficiency of Anti-Stokes Luminescence

Table 13.   Energy Transfer Coefficients, $cm^3 \cdot s^{-1}$

Transfer Phase	Sensitzer-assisted		Via activator ions	
	Forward	Back	Forward	Back
First Transfer of a one-photon excitation	$1 \cdot 10^{-15}$	$5 \cdot 10^{-16}$	$1 \cdot 10^{-16}$	$1 \cdot 10^{-16}$
Second Transfer:				
Red Band	$5 \cdot 10^{-16}$	$3 \cdot 10^{-18}$	$2 \cdot 10^{-17}$	$2 \cdot 10^{-17}$
Green Band	$5 \cdot 10^{-17}$	$5 \cdot 10^{-18}$	$2 \cdot 10^{-17}$	$2 \cdot 10^{-17}$

Table 14. Probabilities of Intra-Center Transitions in s^{-1}

Transition	Ion	Initial Term	Final Term	Transition Probability
Radiative	Yb	$^2F_{5/2}$	$^2F_{7/2}$	$2,5 \cdot 10^3$
	Er	$^4I_{13/2}$	$^4I_{15/2}$	$2 \cdot 10^2$
		$^4I_{11/2}$	$^4I_{15/2}$	$2 \cdot 10^2$
		$^4F_{9/2}$	$^4I_{15/2}$	$1 \cdot 10^4$
		$^4S_{3/2}$	$^4I_{15/2}$	$1 \cdot 10^4$
Multiphonon Relaxation	Er	$^4I_{11/2}$	$^4I_{13/2}$	$3 \cdot 10^2$
		$^4S_{3/2}$	$^4F_{9/2}$	$1 \cdot 10^3$

probabilities of energy transfer and intra-center multiphonon relaxation w_Φ on temperature. We obtained similar data in our work, but will not dwell especially on these calculations since, along with change in the probabilities of the above-mentioned processes, changes in the spectral matching between pumping source and luminophor IR absorption are of major importance in this case. Thus for YOCl luminophors pumped at the long-wavelength end of the excitation spectrum (λ_B=1.06 μm), the value of η grows for heating above room temperature, whereas luminescence efficiency starts to fall off by 100 K for pumping in the center of the absorption band.

The dramatic changes in the anti-Stokes excitation spectra for various temperatures are quite naturally explained by the changes in the ground state Stark component of Yb^{3+} ions and the nonlinearity of anti-Stokes luminescence. In particular this is related to the absolute efficiency of different luminescence bands in YOCl:Yb (10%), Er(10%) luminophors, which were measured in our papers [73,74] at room, nitrogen and helium temperatures for λ_B=1.52 μm and I=10 W/cm^2. The maximum quantum efficiency of $\eta \approx 2\%$ was acheived for the red luminescence band at T=77 K, decreasing (to 0.4%) at room and helium temperatures.

The efficiencies of the green and indigo bands are significantly lower, comparable to one another and dependent on temperature in a similar way (though less extreme). These values for η are in satisfactory agreement with the same set of center parameters as in the case of anti-Stokes pumping of the red and green emission bands in the 0.9 to 1.0 μm range. The coefficient of the first phase of energy cooperation of $2[^4I_{13/2}] \rightarrow {}^4I_{9/2} + {}^4I_{15/2}$, absent in the summing of two excitations, is taken equal to $10^{-16} cm^3 s^{-1}$ in accordance with the discussion presented earlier, and $\sigma_{Er} \approx 1 \cdot 10^{-20} cm^2$ as per Eq. (29). Then for I=10 W/cm^2, C_A=3$\cdot$10^{21}cm^{-3}, the volume pumping density

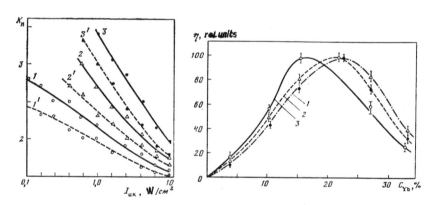

Fig. 28. Dependence of the Degree of Nonlinearity K_H on Pumping Intensity I_{HK} for the Red (1), Green (2), and Blue(2) Emission Bands of YOCl:Yb,Er Luminophors During Pumping by a He-Ne Laser (λ=1.52 μm) at Room (1, 2, 3) and Nitrogen (1', 2', 3') Temperatures

Fig. 29. Normalized Dependence of the Efficiency of Red Band Luminescence in YOCl:Yb,Er Luminophor on Yb Concentration for Two Er Concentrations (1- for C_{Er}=3%, 2- C_{Er}=10%) and additional doping with Dy quenching ions (3- for C_{Dy}=0.1% and C_{Er}=3%)

$\Omega = \sigma C_A I = 2.3 \cdot 10^{21}$photons·cm^{-3}·s^{-1}, which for $\tau_1 = 5 \cdot 10^{-3}$s gives the concentration of singly-pumped Er ions ($^4I_{13/2}$) to be $N_1 = 1 \cdot 10^{19}$cm^{-3}. However, for $P_1 = 10^{-16}$cm^{-3}·s^{-1}, the probability of pumping energy cooperation corresponding to the given concentration of excited centers is $P_1 N_1 = 10^3$s^{-1}, i.e. even twice as high as the probability of intra-center decay. That these processes are of roughly the same probability is attested by the fact that at the indicated pumping intensities the function $J(I_{ИК})$ is already approaching $I_{ИК}^{3/2}$. Thus in this regime it is necessary to solve a quadratic equation for N_1 with consideration of cooperation processes:

$$N_1/\tau_1 + 2P_1 N_1^2 = \sigma C_A I \equiv \Omega \tag{69}$$

which, for substituting the values of τ_1 and P_1 given above and $\Omega = 2.3 \cdot 10^{21}$ photons·cm^{-3}·s^{-1} gives $N_1 = 2.9 \cdot 10^{18}$cm^{-3}. Considering that Er^{3+} very quickly relaxes from the "doubly" pumped state to $^4I_{11/2}$, and furthermore that roughly half the excitation is transferred to Yb^{3+}, during which the states $^2F_{5/2}$(Yb^{3+}) and $^4I_{11/2}$(Er^{3+}) have a common lifetime $\bar\tau = 4.3 \cdot 10^{-4}$s, we obtain

$$N_2[\text{Er}(^4I_{11/2})] = N_S[\text{Yb}(^2F_{5/2})] = 2 \cdot 10^{-17}\text{cm}^{-3}. \tag{70}$$

Thereupon the number of acts of pumping energy cooperation of Er^{3+} ions ($^4I_{13/2}$) and Yb^{3+} ions ($^2F_{5/2}$) for a transfer coefficient of $P_{2s} = 3 \cdot 10^{-16}$cm^3·s^{-1} is equal to $1.8 \cdot 10^{20}$cm^{-3}·s^{-1}. Figuring in the quantum luminescence yield during Stokes pumping (0.25) this gives a volume anti-Stokes transition density of $4 \cdot 10^{19}$cm^{-3}·s^{-1}. The ratio of this latter to the volume pumping density corresponds to a quantum yield of 2%, which is only a few times higher than the experimental value for red luminescence quantum efficiency. Considering how close this estimate is, and the unavoidable additional energy losses (by reflection of pumping radiation, weaker excitation of underlying layers, and reabsorption of the visible luminescence), we may consider these results to be in quite satisfactory agreement.

4. COMPARISON OF VARIOUS CLASSES OF LUMINOPHORS WHICH MAKE IR RADIATION VISIBLE

In addition to the cooperative luminophors examined in some detail in this work, there are also other luminophors, some of which are rare-earth activated, which are used for making IR radiation visible. It is interesting to compare the properties of these luminophors with one another in order to clarify where cooperative luminophors stand in this group.

All IR-sensitive luminophors, with the exception of the cooperatives, require supplementary shorter-wavelength excitation (usually ultraviolet) which may be in the form of a preliminary or a stationary excitation. It is for this reason that they, unlike the cooperative luminophors, are unsuitable for developing GaAs diode-based visible-band LEDs. (The design of such LEDs was described in a series of articles [94-98,76] and we will not dwell on that here). On the other hand, luminophors with recombination luminescence mechanisms provide much higher sensitivity and significantly broader spectral range for recording IR radiation fields (including from lasers) than do luminophors with cooperative luminescence mechanisms. Therefore, despite the necessity of supplementary irradiation, they are found in a whole series of applications.

CaS SrS:Eu,Sm and CaS SrS:Cl,Sm are classic examples of luminophors whose afterglow is effectively stimulated by IR (in the 0.7 to 1.5 μm region) [99,100]. These luminophors store up a sizeable light sum during ultraviolet and even indigo-band pumping, which they maintain for long periods of time after the cessation of pumping. The amount of this light sum is determined by the concentration of localized nonequilibrium electrons in trapping centers and holes in luminescence centers. The levels of these centers are so deep that room temperature thermal liberation of charges practically does not take place. During the excitation process, the nonequilibrium charge carriers may populate more than half of these centers and maintain that population for several days. A screen made of this luminophor can luminesce up to 10^{15}photons/cm^2 upon pumping as much as a week later, as long as it is irradiated by an IR beam wherein each of the photons in the flux carries an energy much less than for luminescence excitation. The action of these photons results in liberation of the nonequilibrium charge carriers stored earlier by the action of the more energetic exitation phonons, which leads to a gain in the radiative recombination flux and a corresponding gain in luminescence brightness. In this way we manage to circumvent the Stokes rule with the help of recombination luminophors and indirectly carry out the conversion of IR radiation to visible light. The efficiency of such conversion in these luminophors is over 1%. This efficiency is attained in cooperative luminophors only at high pumping intensities I ($I \approx 1$ W/cm^2). This is more easily attained in recombination luminophors at lower pumping densities, and rare-earth luminophors are no exception here.

Roughly the same quantum efficiency was attained for converting IR beams to visible light during the developement of two-activator zincblende luminophors of the ZnS:Cu,Me,Cl type, where Me is Sm, Pb or Mn. The preliminary ultraviolet irradiation in these luminophors results in charge transfer of the "copper" centers, and the action of the IR photons liberates the holes from these centers with subsequent radiative recombination on the second activator, which defines the scintillation spectrum. Thus, in the case of Sm^{3+}, distinct narrow bands in the yellow-orange region appear in the spectrum, which are easily identified as transitions between known terms of this ion [101,102] (Figs. 30,31). One of the fundamental advantages of zinc sulfide luminophors over luminophors with alkali-earth bases is their higher chemical and radiation stability. The IR threshhold density corresponding to a 10% excess over the phonon afterglow level (0.1 to 0.01 nt), attained after a time period of from a few dozen seconds to minutes, is ~10^{-6} W/cm^2 for all these luminophors. This value is roughly 3 orders of magnitude better than for cooperative luminophors at maximum sensitivity. Actually, due to the nonlinearity of the excitation summing mechanism in cooperative luminophors, their emission so quickly diminishes upon reduction of pumping density that, from our observations, even in a darkened room the IR threshhold density is no less than 1 mW/cm^2. This value is in satisfactory agreement with data presented earlier on the absolute efficiency of anti-Stokes luminescence. Actually, if for an IR density of 1 W/cm^2 the conversion efficiency of the best cooperative luminophors reaches 1%, then for $I = 1$ mW/cm^2 it is only $1 \cdot 10^{-3}$%. Correspondingly, the luminescence power intensity will be 10^{-8} W/cm^2. This means that, even for the green band, to which the eye is more sensitive (the optical equivalent power for green light is, as is well-known, 680 lm/W), the luminescence brightness will be only several hundred nit. At such low brightness the contrast threshhold and resolving power of the eye are already far from optimal. Thus the use of cooperative luminophors is inadvisable for low IR radiation densities.

At low IR intensities scintillating luminophors provide significantly higher brightness of visual images in the investigated radiation fields than do cooperative luminophors. However their limited light sum storage results in the fact that, during steady-state action by IR beams the scintillation brightness begins to subside after some time, and the higher the IR density, the quicker the subsidence. Thus, according to approximate calculations that satisfactorily agree with experiment, for the efficiency mentioned above (1%) to a light sum of 10^{15}photons/cm^2 and IR density $1 \cdot 10^{-6}$W/cm^2 (i.e., $5 \cdot 10^{12}$photons$\cdot$cm$^{-2} \cdot$s^{-1}), de-excitation begins to appear only after several minutes, but for $I = 10^{-3}$W/cm^2 it appears after only tenths of a second. In the latter case it is obviously difficult to get even a qualitative representation of the structure of the IR image, let alone a quantitative treatment, since high radiation density images may become less bright than those on which no IR is incident. This effect may be obviated in part by using a rotating luminescent screen (e.g., a cylinder), with part of its surface

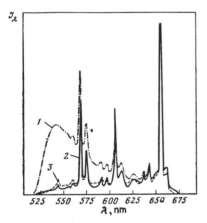 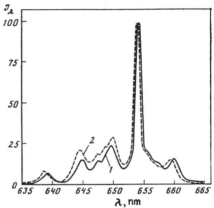

Fig. 30. Emission Spectrum of ZnS:Cu,Sm Luminophors at Various Pumping Conditions

1- DRT-220 ultraviolet lamp (with filter passing λ=365 nm radiation); 2- Scintillation under the action of GaAs laser radiation (λ=916 nm); 3- Afterglow pumped by LGI-21 laser (λ=337 nm)

Fig. 31. Scintillation Spectrum of the Red Band of Sm^{3+} Ions ($^4G_{5/2} \rightarrow {}^6H_{9/2}$ transition) Under the Action of GaAs Laser Radiation (λ=916 nm)

1- ZnS:Cu,Sm; 2- ZnS:Ag,Sm

Fig. 32. Normalized Spectum of Luminophor IR Sensitivity

1- ZnS:Cu,Co; 2- ZnS:Cu,Mn; 3- ZnS:Cu,Sm; 4- ZnS:Cu,Pb; ZnS:Ag,Sm

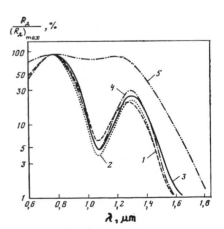

being constantly pumped by the ultraviolet and IR images being projected on the other part of the screen. The visible scintillation is maintained constant for a practically unlimited time, and its value remains proportional to the IR flux density all the way down to 10^{-2} W/cm^2, attaining several hundred nit at this density (corresponding to television standards). By this very method we can get 100 to 200-times better images of the radiating surfaces of GaAs lasers at their natural frequency ($\lambda = 0.91$ μm) and with this to verify the uniformity of radiation from a p-n junction [103]. These screens are also sensitive to those long wavelengths lying beyond the photocathode sensitivity threshhold in opto-electronic converters but, on the other hand, lying within the transparency region of widely-used semiconductor materials such as Si or Ge. Thus these screens may be used to advantage in IR-laser barium vapor projection microscopes [104-107] for monitoring the internal structure of various parts made of these materials (in particular, integrated circuits). At the same time, since the average power of laser radiation striking the barium vapors reaches tens of watts and the wavelength (1.52 μm) falls within the sensitivity of cooperative luminophors, then for image dimensions of a few square centimeters it is better to use screens made from cooperative luminophors. At those pumping densities their efficiency reaches 1%, which ensures high brightness (up to 10 nt), and a cubic dependence of emission brightness on the pumping density, which arises from the summing of three excitations, ensures high image contrast. If images of larger area (e.g., 1 m^2) are needed, then in that case scintillation luminophors work better.

One of the fundamental shortcomings of cooperative luminophors is their narrow sensitivity range, which proceeds directly from the mechanism by which they operate (cf. Section 2); what is more, there are only two basic sensitivity bands known at the present time. Optically sensitive luminophors have sensitivity extending down to 2 μm at room temperature, and they may operate also at 3 to 4 μm when the screen is cooled to nitrogen temperature.

Temperature-sensitive luminophor screens have incomparably wider sensitivity zones. Their sensitivity range is determined not only by the absorption spectrum of the luminophor itself, but also the substrate, on which a special absorbing layer may be deposited. For example, screens on a lavsan substrate provided with a layer of aluminum 10 to 50 Å thick have a sensitivity region extending from the near-IR to UHF radio waves [108]. The operating mechanism of these luminophors, basic parameters of the screens that go with them, and their range of application are covered in a series of publications by our colleagues at the Lebedev Physics Institute [FIAN][108-111]. Therefore we will only make note of the fact here that their threshhold for visual registration is roughly the same as that for anti-Stokes luminophors. However, they have in principle poorer spatial and temporal resolution by dint of thermal mechanisms. Thus their application to the near-IR is inadvisable.

This is an appropriate place to mention yet another class of luminophors sensitive to IR radiation. That would be ZnS:Cu,Co,Cl and ZnS:Cu,Ni,Cl [109,112] which are quenched by IR action (1.3 to 1.5 μm). Although these luminophors are inferior to the scintillators in terms of sensitivity and delay time, they possess one overriding advantage: on screens made from them, a negative of the image is generated which stays on that screen for something on the order of minutes at room temperature. During this period of time it is possible to examine it in detail, or print it on photosensitive paper, obtaining a positive image. The fundamental properties of various classes of IR-sensitive luminophors is presented in Table 15, from which the best may be selected for deciding on a type of screen for a given purpose.

Table 15. Basic Specifications of Luminescent Screens Designed for Visualizing IR Images

Operating Principle		Luminophor	Sensitivity Range, μm	Recording Ths'ld, W/cm^2	Delay Time, s	Resolution mm^{-1}
Energy Summing	Two Excited RE^{3+} Ions	YOCl : Yb, Er	0,9—1,1	10^{-3}	10^{-3}—10^{-4}	15—20
		NaYF$_4$: Yb, Er	0,9—1,1	10^{-3}	10^{-3}—10^{-4}	15—20
		Y$_2$O$_2$S : Yb, Er	0,9—1,1	10^{-3}	10^{-3}—10^{-4}	15—20
	Three Excited RE^{3+} Ions	YOCl : Yb, Er	1,4—1,6	10^{-2}	10^{-2}—10^{-3}	20—25
		BaYF$_5$: Er	1,4—1,6	10^{-2}	10^{-2}—10^{-3}	20—25
Optical Effects	Stimulation	CaS.SrS : Eu, Sm	0,75—1,5	10^{-6}	10^{-1}—10^{-3}	10—15
		ZnS : Cu, Pb, Cl	0,75—1,5	10^{-6}	10^{-1}—10^{-3}	10—15
		ZnS : Cu, Sm, Cl	0,75—1,7	10^{-6}	10^{-1}—10^{-3}	10—15
		ZnS : Fe	0,75—3,5	10^{-5}	10^{-1}—10^{-3}	10—15
	Quenching	ZnS : Cu, Co, Cl	0,75—1,7	10^{-4}—10^{-5}	до 10^3	10
	Thermal Quenching	ZnS.CdS : Ag, Ni, Cl	1—10^5	10^{-3}	1	4—5

CONCLUSIONS

From what has been said, it is obvious that the simplified theory developed here quite fully reflects the spectral and energy characteristics of cooperative luminophors. We will briefly summarize the basic conclusions from this theory, which may be applied in the development of anti-Stokes luminophors and their applications.

1. The efficiency of anti-Stokes luminescence η is limited in principle by the interrelationship of the probabilities of forward and back transitions, i.e. the processes of generation and decay of electronic excitations in RE^{3+} ions. Similar to laser media at the first stage of generating one-photon excitations, these interrelationships are defined by the fundamental ratios between the Einstein coefficients for the probability of optical transitions. In real systems the situation is complicated somewhat by the existence of Stark components and multiple radiative channels, as well as radiationless transitions which, however, do not change the limiting value of η and may be computed based on analysis of the experimental results. In the final stages of cooperation of two or more electronic excitations, the coefficients of summing and of cross-relaxation decay of high-energy electronic excitations during resonant energy transfer are directly equal to one another. Thus we require in principle that the mismatch in energy between transitions participating in energy transfer be covered by lattice phonons in efficienct cooperative luminophors.

63

2. The action of sensitzers in cooperative luminophors basically results in an increase of the ratio of the probability of excitation energy cooperation to the cross-relaxation probability which, in part, allows us to employ higher concentrations of participating ions. It is this circumstance that explains the substantial (by as much as a factor of 10) increase in the efficiency of the red emission of Er^{3+} during IR pumping with $\lambda = 1.52$ μm, hence YOCl:Er luminophors are still prepared with Yb^{3+} ion admixtures, even though these ions do not themselves absorb the pumping radiation.

3. The various colors of the dominant anti-Stokes luminescence of Er^{3+} ions in different bases is chiefly explained by differences in the ratio of energy mismatch between cooperation and inter-ion cross-relaxation of pumping energy, and the phonon frequency in the given lattice. Most important for the red emission band is the number of phonons required for cooperation of pumping energy, while for the green band it is the number of phonons required for the primary cross-relaxation process in the presence of Yb.

4. The absolute value of the quantum yield of present-day anti-Stokes luminophors ($1 < \eta < 10\%$) at the given IR pumping density ($I = 1$ W/cm²) differs by no more than an order of magnitude from the limiting value for the idealized model. The thing that makes RE^{3+} ions unique for cooperation of pumping energy is not the significantly longer lifetime of their excited states, but the small energy width of the corresponding terms, which at the same time limits the possibility spectral matching with different sources of IR radiation.

5. The use of anti-Stokes luminophors for recording fields of IR laser radiation would only be warranted for rather high densities of this radiation ($I > 10^{-2}$W/cm²); it would be more convenient to achieve the same thing by using scintillation luminophors with recombination luminescence mechanisms at low values of I.

6. Cooperative luminophors are unsuitable for increasing the luminous yield of incandescent lamps, since due to their narrow range of sensitivity — and that with low efficiency — they can only use an insignificant part of the lamp filament's IR radiation.

The authors are very grateful to S. A. Fridman, E. Ya. Arapova and V. V. Shchayenko for synthesizing the luminophors we used, as well as A. A. Glushko, N. N. Sibel'din and A. A. Kut'yenkov for taking part in the measurement of their characteristics.

BIBLIOGRAPHY

1. Ovsyankin, V. V. and Feofilov, P. P., "Cooperative Processes in Luminescing Systems", IZVESTIYA AN SSSR, SER. FIZ., Vol 37 pp 262-272, 1973

2. Auzel, F. E., "Materials and Structures Using Anti-Stokes Luminophors with Energy Transfer", TIIER [Russian translation of TRANS. IEEE] Vol 61 pp 87-120, 1973

3. Chukova, Yu. P., "Antistoksova lyuminestsentsiya i novyye

vozmozhnosti yeyë primeneniy" [Anti-Stokes Luminescence and New Application Possibilities For It], Moscow, Sovetskoye Radio, 1980, 193 pp

4. Bloembergen, N., "Solid-State Infrared Quantum Counters", PHYS. REV. LETT. Vol 2 pp 84-85, 1959

5. Ovsyankin, V. V. and Feofilov, P. P., "Cooperative Sensitization of Luminescence in Rare-Earth Activated Crystals", PIS'MA V ZhETF Vol 4 pp 471-474, 1966

6. Ovsyankin, V. V. and Feofilov, P. P., "On the Mechanism of Summing Electronic Excitations in Activated Crystals", PIS'MA V ZhETF Vol 3 494-497, 1966

7. Auzel, F., "Compteur quantique par transfer d'energie entre deux ions de terres rares dans un tungstate mixte et dans un verre", C. R. ACAD. SCI. B Vol 262 pp 1016-1019, 1966

8. Auzel, F., "Compteur quantique par transfer d'energie de Yb^{3+}, Tm^{3+} dans un tungstate mixte et dans un verre germanate", C. R. ACAD. SCI. B Vol 263 pp 819-821, 1966

9. Brown, M. R. and Shand, W. A., "Infrared Quantum Counter Action in Er-Doped Fluoride Lattice", PHYS. REV. LETT. Vol 12 pp 367-369, 1964

10. Bakumenko, V. L., Vlasov, A. N., Kovarskaya, Ye. S. et al, "On the Step-wise Excitation of Luminescence in Er^{3+}-Activated $CaWO_4$", PIS'MA V ZhETF Vol 2 pp 27-30, 1965

11. Miyakawa, T. and Dexter, D. L., "Phonon Sidebands, Multiphonon Relaxation of Excited States and Phonon-Assisted Energy Transfer Between Ions in Solids", PHYS. REV. B - SOLID STATE Vol 1 pp 2961-2969, 1970

12. Miyakawa, T. and Dexter, D. L., "Cooperative and Stepwise Excitation of Luminescence of Trivalent Rare Earth Ions in Yb^{3+}-Sensitized Crystals", PHYS. REV. B - SOLID STATE Vol 1 pp 70-80, 1970

13. Förster, Th., "Zwischenmole kulare Energiewanderung und Fluoreeczens" ANN. PHYS. Vol 2 pp 55-75, 1948

14. Dexter, D. L., "A Theory of Sensitized Luminescence in Solids", J. CHEM. PHYS. Vol 21 pp 835-850, 1953

15. Galanin, M. D., "Resonant Transfer of Excitation Energy in Luminescing Solutions", TRUDY FIAN SSSR Vol 12 pp 3-35, 1960

16. Agranovich, V. M. and Galanin, M. D., "Perenos energii elektronnogo vozbuzhdeniya v kondensirovannykh sredakh" [Transfer of Electronic Excitation Energy in Condensed Matter], Moscow, Nauka, 1978, 383 pp

17. Livanova, L. D., Saytkulov, I. G. and Stolov, A. L., "Processes of Quantum Summing in Tb^{3+}- and Yb^{3+}-Activated CaF_2 and SrF_2 Single

Crystals", FTT Vol 11 pp 918-923, 1969

18. Hohnson, L. F. and Guggenheim, J. J., "Infrared Pumped Visible Laser", APPL. PHYS. LETT. Vol 19 pp 44-47, 1971

19. Antipenko, B. M., Mak, A. A., Raba, O. B. et al, "Stepwise Sensitization Mechanism in $^4F_{9/2} \to {}^4I_{11/2}$ and $^4F_{9/2} \to {}^4I_{13/2}$ Laser Transitions of Er^{3+} Ions in $BaYb_2F_8$ Crystals", KVANTOVAYA ELEKTRONIKA Vol 9 pp 1614-1619, 1982

20. Ovsyankin, V. V. and Feofilov, P. P., "Cooperative Sensitization of Photomechanical and Photochemical Proceses" in the book "Molekulyarnaya fotonika" [Molecular Photonics], Leningrad, Nauka, 1970

21. Kushida, T., "Probability of Energy Transfer and Cooperative Transitions in Rare-Earth Ions in the Solid State", IZVESTIYA AN SSSR, SER. FIZ., Vol 37 pp 273-283, 1972

22. Kushida, T., "Energy Transfer and Cooperative Optical Transition on Rare-Earth Doped Inorganic Material", J. PHYS. SOC. JAPAN Vol 4 pp 1318-1337, 1973

23. Ganrud, W. B. and Moos, H. W., J. CHEM. PHYS. Vol 49 pp 2170-2173, 1973

24. Van der Ziel, J. P., van Uitert, L. G. and Grodkiewisc, W. H., "Factors Controlling Infrared-Pumped Visible Emission of $Yb^{3+}-Er^{3+}$ in the Scheelites", J. APPL. PHYS. Vol 41 pp 3308-3315, 1970

25. Hewes, R. A. and Sarver, J. F., "Infrared Excitation Processes for the Visible Luminescence of Er^{3+}, Ho^{3+} and Tm^{3+} in Yb^{3+}-Sensitized Rare-Earth Trifluorides", PHYS. REV. Vol 182 pp 427-439, 1962

26. Mita, Y., "Luminescence Processes in Yb^{3+}-Sensitized Rare-Earth Phosphors", J. APPL. PHYS. Vol 43 pp 1772-1778, 1972

27. Sergeyev, I. I., "Temperature Dependence of Anti-Stokes Luminescence in YF_3:Yb,Er and YOCl:Yb,Er Systems", "Preprint Instituta fiziki AN BSSR No 125", Minsk, 1977, 53pp

28. Sergeyev, I. I. and Kuznetsova, V. V., "Temperature Dependence of Anti-Stokes Luminescence in YOCl:Yb,Er Systems", ZhPS Vol27 pp423-428, 1977

29. Sergeyev, I. I. and Kuznetsova, V. V., "Investigation of Limits to IR-to-Visible Radiation Conversion Yield in YF_3:Yb,Er", ZhPS Vol 29 pp 470-478, 1978

30. Sergeyev, I. I. and Kuznetsova, V. V., "Numerical Analysis of Concentration Effects in Anti-Stokes Luminescence of Rare-Earth Luminophors", IZVESTIYA AN BSSR, SER. FIZ.-MAT. NAUK, Vol 5 pp71-75, 1980

31. Sergeyev, I. I., "Issledovaniye antistoksovoy lyuminetsentsii ftoridov RZE veroyatnostnym metodom" [Investigation of Anti-Stokes Luminescence in RZE Fluoride Using the Probability Method], Dissertation for Candidate of Phys.-Math. Sci. Degree, Minsk, Inst. fiziki AN BSSR, 1980, 171 pp

32. Sergeyev, I. I. and Kuznetsova, V. V., "Anti-Stokes Luminescence of Erbium in Fluorides and Oxychlorides of Yttrium Pumped in the 1.5 μm Range. A Numerical Analysis" in "Preprint Instituta fiziki AN BSSR No 252", Minsk, 1981, 45 pp

33. Morgenshtern, Z. L. and Neustroyev, V. B., "Luminescence Yield of Ruby Under Resonant Pumping", OPTIKA I SPEKTROSKOPIYA Vol 32 pp 953-958, 1972

34. Dianov, Ye. M., Karasik, A. Ya., Neustoyev, V. B. et al, "Direct Measurements of Luminescence Quantum Yield From the Metastable State $^4F_{3/2}$ of Nd^{3+} in $Y_3Al_5O_{12}$", DAN SSSR Vol 224 pp 64-67, 1975

35. Avanesov, A. G., Voron'ko, Yu. K., Denker, B. I. et al, "Measurements of Absolute Luminescence Quantum Yield of Neodymium in Highly-Concentrated Glasses Activated With Chromium", KVANTOVAYA ELEKTRONIKA Vol 6 pp 2253-2256, 1979

36. Galanin, M. D., Kut'yenkov, A. A., Smorchkov, V. N. at al, "Measurements of the Quantum Yield of Photoluminescence in Red Dyes by the Vavilov Method and the Integrating Sphere Method", OPTIKA I SPEKTROSKOPIYA Vol 53, pp 683-689, 1982

37. Kingsley, J. D., "Analysis of Energy Transfer and Infrared-to-Visible Conversion in LaF_3:Yb,Er", J. APPL. PHYS. Vol 11 pp 175-182, 1970

38. Ivlev, Yu. A., "Towards a Calculation of the Efficiency of Cooperative Luminescence in the Steady State", OPTIKA I SPEKTROSKOPIYA Vol 28 pp 804-805, 1970

39. Ivlev, Yu. A., "Towards a Calculation of Cooperative Luminescence for Three Impurity Atoms in the Crystal", OPTIKA I SPEKTROSKOPIYA Vol 30 pp721-729, 1971

40. Timofeyev, Yu. P., Fok, M. V. and Fridman, S. A., "Luminescence Method of Visualization of Invisible Images" IZVESTIYA AN SSSR, SER. FIZ., Vol 43 pp 1203-1211, 1979

41. Timofeyev, Y. P. and Fok, M. V., "Kinetics of Recombinant Interactions of Impurity Centers in Crystal Phosphors", TRUDY FIAN SSSR Vol 117 pp 3-54, 1980

42. Grant, W. J. C., "Role of Rate Equations in the Theory of Luminescence Energy Transfer", PHYS. REV. B - SOLID STATE Vol 4 pp 648-663, 1971

43. Agabekyan, A. S. and Memekyan, A. O., "On Application Criteria for

tyhe Theory of Resonant Energy Transfer", OPTIKA I SPEKTROSKOPIYA Vol 32 p 288-295, 1972

44. Zveryev, M. G., Kurateyev, I. I., Myshlyayev, N. F. and Onishchenko, A. M., "Kinetics of IR-Pumped Visible Luminescence of Er^{3+} Ions in La_2O_2S Activated With Yb^{3+} and Er^{3+} ", KVANTOVAYA ELEKTRONIKA Vol 4 pp 866-871, 1977

45. Glushko, A. A., Osiko, V. V., Timofeyev, Yu. P. and Shcherbakov, I. A., "Kinetics of Population and Decay in High Energy Excited States of RE^{3+} Ions Assuming Highly Incoherent Interactions in Intermediate States", ZhETF Vol 79 pp 194-206, 1980

46. Shcherbakov, I. A., "Investigation of Excitation Energy Relaxation Processes in Crystals and Glasses Activated With Rare-Earth Element Ions", Dissertation for Doctor of Phys.-Math. Sci. Degree, Moscow, Fiz. Inst. imeni P. N. Lebedeva AN SSSR, 1978, 357 pp

47. Lyashko, O. M., Khomenko, V. S. and Kuznetsova, V. V., "Energy Levels of Impurity Ions Er^{3+} and Yb^{3+} in Gadolinium Oxychloride and Their UV- and IR-Induced Luminescence", VESTNIK AN BSSR. FIZ.-MAT. NAUKI Vol 6 pp 123-127, 1979

48. Antipenko, B. M. and Nikolayev, V. B., "On the Role of Diffusion of Excitations in the Cooperative Sensitization Process", OPTIKA I SPEKTROSKOPIYA Vol 39 pp 290-295, 1975

49. Rich, T. C. and Pinnow, D. A., "Exploring the Ultimate Efficiency in Infrared-to-Visible Converting Phosphors Activated With Er and Sensitized With Yb", J. APPL. PHYS. Vol 43 pp 2357-2365, 1972

50. Arapova, E. Ya., Baryshnikov, N. V., Derevyanko, A. S. et al,"Anti-Stokes Luminescence of Er^{3+}-Yb^{3+} Pairs in Oxychloride and Fluoride Bases", IZVESTIYA AN SSSR, SER. FIZ., Vol 38 pp 1185-1189, 1974

51. Kaminskiy, A. A., "Lazernyye kristally" [Laser Crystals], Moscow, Nauka, 1975, 256 pp

52. Zhabotinskiy, M. Ye.,[editor] "Lazernyye fosfatnyye stekla" [Laser Phosphate Glasses], Moscow, Nauka, 1980, 352 pp

53. Morozova, V. N., Timofeyev, Yu. P., Fridman, S. A. and Chukova, Yu. P., "Optical Properties of Layers of Anti-Stokes Luminophors", SVETOTEKHNIKA Vol 6 pp 18-20, 1976

54. Feofilov, P. P., "Cooperative Optical Phenomena in Activated Crystals" in the book "Fizika primesnykh tsentrov v kristallakh" [Physics of Impurity Centers in Crystals], Tallinn, Izd-vo AN ESSR, 1972

55. Weber, M. J., "Probabilities for Radiative and Non-Radiative Decay of Er^{3+} in LaF_3", PHYS. REV. Vol 157 pp 262-272, 1967

56. Weber, M. J., "Radiative Multiphonon Relaxation of Rare-Earth Ions in

Y_2O_3", PHYS. REV. Vol 171 pp 283-291, 1968

57. Partlow, W. D. and Moos, H. W., "Multiphonon Relaxation of $LaCl_3:Nd^{3+}$", PHYS. REV. Vol 157 pp 252-257, 1967

58. Risberg, L. A. and Moos, H. W.,"Multiphonon Orbit-Lattice Relaxation of Excited States of Rare-Earth Ions in Crystals", PHYS. REV. Vol 174 pp 429-438, 1968

59. Kuroda, H., Shionoga, Sh. and Kushida, T., "Mechanism ans Controlling Factors of Infrared-to-Visible Conversion Process in Er^{3+} and Yb^{3+} Doped Phosphors", J. PHYS. SOC. JAPAN Vol 33 pp 125-141, 1972

60. Yamada, N., Shionoga, Sh. and Kushida, T., "Phonon-Assisted Energy Transfer Between Rare-Earth Ions", J. PHYS. SOC. JAPAN Vol 32 pp 1577-1586, 1972

61. Nuzel, F., "Multiphonon-Assisted Anti-Stokes and Stokes Fluorescence of Triply Ionized Rare-Earth Ions", PHYS. REV. B - SOLID STATE Vol 13 pp 2809-2817, 1976

62. Perlin, Yu. V., "Contemporary Methods in the Theory of Multiphonon Processes", USPEKHI FIZ. NAUK Vol 80 pp 553-595, 1963

63. Gamurar', V. Ya., Perlin, Yu. V. and Pukerblat, B. S., "Multiphonon Radiationless Relaxation in Rare-Earth Impurity Ions", IZVESTIYA AN SSSR, SER. FIZ., Vol 35 pp 1429-1432, 1971

64. Gyërlikh, P., Karras, Kh., Ketitu, K. and Leman, R., "Spektroskopicheskiye svoystva aktivirovannykh lazernykh kristallov" [Spectroscopic Properties of Activated Laser Crystals], Moscow, Nauka, 1966, 207 pp

65. Matsubara, T., "A Proposed Method for Predicting Emission Color in Er^{3+} and Yb^{3+} Doped Phosphors", JAPAN. J. APPL. PHYS. Vol 11 pp 1579-1580, 1972

66. Garlick, G. J. F., "Radiative and Nonradiative Transition of Erbium Ions Dispersed in Solids", APPL. PHYS. LETT. Vol 47A pp 441-442, 1974

67. Zhekov, V. I., "Spectral and Lasing Properties of Yttrium-Erbium-Aluminum Garnet Crystals", Dissertation for Candidate of Phys.-Math. Sci. Degree, Moscow, Fiz. Inst. im.P. N. Lebedeva AN SSSR, 1977, 153 pp

68. Glushko, A. A., Osiko, V. V., Timofeyev, Yu. P. and Shcherbakov, I. A., "Effect of Crystal Lattice on Processes Populating Higher Excited States of RE^{3+} Ions During Infrared Pumping", ZhETF Vol 78 pp 53-61, 1980

69. Ovsyankin, V. V. and Feofilov, P. P., "Triple Optical Resonance in BaF_2-Er^{3+} Crystals", OPTIKA I SPEKTROSKOPIYA Vol 20 pp 526-528, 1966

70. Veber, C. M., "Infrared to Visible Conversions in CaF_2-Er^{3+}: A

Sequential-Pair Process" J. APPL. PHYS. Vol 44 pp 3263-3265, 1973

71. Fridman, S. A., Arapova, E. Ya., Mitrofanova, N. V. et al, "Luminescence Methods for Visualization of Long Wavelength Radiation", IZVESTIYA AN SSSR, SER. FIZ., Vol 37 pp 783-789, 1973

72. Hewes, R. A., "Multiphonon Excitation and Efficiency in the Yb^{3+}-RE^{3+} (Ho^{3+}, Er^{3+}, Tm^{3+}) Systems", J. LUMINESCENCE Vol 1 pp778-796, 1970

73. Arapova, E. Ya., Zamkovets, N. V. and Sibel'din, N. N., "Anti-Stokes Luminescence of $YOCl$-Yb^{3+},Er^{3+} Under Laser Pumping in the 1.5 μm Range", OPTIKA I SPEKTROSKOPIYA Vol 41 pp 890-891, 1976

74. Johnson, L. F., Geusic, J. E., Guggenheim, H. J. et al, "Comments on Materials for Efficient Infrared Conversion", APPL. PHYS. LETT. Vol 15 pp 48-50, 1969

75. Arapova, E. Ya., Baryshnikov, N. V., Bochkarev, E. P. et al, "LEDs Based on Anti-Stokes Luminophors and Semiconductor IR Radiation Sources" in the book "Materialy Vsesoyuz. konf. 'Primeneniye elektro-lyuminestsentsii v narodnom khozyaystve' " [Proceedings of the All-Union Conference "Applications of Electroluminescence in the Economy"] Chernovtsy, 1971, 86 pp

76. Van Uitert, L. G., Singh, S., Levinstein, H. J. et al, "Efficient Infrared-to-Visible Conversion by Rare-Earth Oxychlorides", APPL. PHYS. LETT. Vol 15 pp 53-54, 1969

77. Johnson, L. F., Guggenheim, H. T., Rich, T. C. and Ostermayer, F. W., "Infrared-to-Visible Conversion by Rare-Earth Ions in Crystals", J. APPL. PHYS. Vol 43 pp 1125-1137, 1972

78. Yocom, P. N., Witthe, J. P. and Ladany, "Rare-Earth-Doped Oxysulfides for GaAs-Pumped Luminescent Devices", MET. TRANS. Vol 2 pp763-767, 1971

79. Arapova, E. Ya., Glushko, A. A. and Timofeyev, Yu. P., "Spectrum of IR-Induced Visible Light in Yb^{3+}- and Er^{3+}-Activated Luminophors", ZhPS Vol 29 pp 876-881, 1978

80. De Hair, J. Th. W., "The Intermediate Role of Cd^{3+} in Energy Transfer From Sensitizer to Activator", J. LUMINESCENCE Vol 18/19 pp 797-800, 1979

81. Kato, T., Yamomoto, H. and Otomo, Y., "$NaLnF_4$-Yb^{3+},Er^{3+} (Ln=Y, Gd, La) Efficient Green Emitting Infrared Excited Phosphors", J. ELECTRON. SOC. Vol 119, pp 1561-1564, 1972

82. Menyuk, N., Dwight, K. and Pierce, J. W., "$NaYF_4$-Ab,Er, An Efficient Up-Conversion Phosphor", APPL. PHYS. LETT. Vol 21 pp 159-161, 1972

83. Sommerdijk, J. L., "Influence of Host Lattice on the Infrared-Excited Visible Luminescence in Yb^{3+}, Er^{3+}-Doped Fluorides", J.

LUMINESCENCE Vol 6 pp 61-67, 1973

84. Brown, M. R., Roots, K. G. and Shand, W. A., "Energy Levels of Er^{3+} in $LiYf_4$", PHYS. REV. B - SOLID STATE Vol 2 pp 593-602, 1969

85. Seki, Y. and Furukawa, Y., "Improved Conversion Efficiency of Infrared Stimulable Phosphor $(Y_{0.6-x}La_x)_3 \cdot OCl_7 Yb^{3+},Er^{3+}$", JAPAN. J. APPL. PHYS. Vol 10 pp 1293-1294, 1971

86. Fok, M. V., "Investigation of Afterglow of Eu^{3+} Ions in Thorium Oxide-Based Phosphors", OPTIKA I SPEKTROSKOPIYA Vol 2 pp 127-135, 1957

87. Glushko, A. A., "Investigation of Anti-Stokes Radiation in Crystals Activated With Yttrium and Erbium Ions", Dissertation for Candidate of Phys.-Math. Sci., Moscow, Fiz. Inst. imeni P. N. Lebedeva AN SSSR, 1980, 117 pp

88. Geusic, J. E., Ostermayer, F. W. and Marcos, L. G., "Efficiency of Red, Green and Blue Infrared-to Visible Conversion Sources", J. APPL. PHYS. Vol 42 pp 1958-1960, 1971

89. Chukova, Yu. P., "Thermodynamic Limits to the Efficiency of Anti-Stokes Luminophors", ZhPS Vol 20 pp 412-416, 1974

90. Pasternak, J., "Thermodynamical Considerations of the Quantum Efficiency of Anti-Stokes Cooperative Luminescence", J. LUMINESCENCE Vol 9 pp 249-256, 1974

91. Kazaryan, A. K., Timofeyev, Yu. P. and Fok, M. V., "Kinetic Limits to the Efficiency of Anti-Stokes Luminescence in RE^{3+} Ions", KRAT. SOOBShch. PO FIZIKE FIAN Vol 3 pp 51-55, 1983

92. Barnet, A. M. and Kh'yuman, F. K., "Chromatic Semiconductor Alphanumeric Indicator", ELEKTRONIKA Vol 10 pp 3-7, 1970

93. Berg, A. and Dean, P., "Svetodiody" [Light-Emitting Diodes], translated from English, A. E. Yunovich, ed., Moscow, Mir, 1973, 186 pp

94. Neytgauz, L. M., Nosov, Yu. P., Boyev, E. I. et al, "Light Sources Based on Gallium Arsenide IR Emitters and Anti-Stokes Luminophors", ELEKTRON. TEKHNIKA. SER. POLUPROVODNIKOVYYE PRIBORY Vol 2 pp 81-87, 1974

95. Mirakova, M. G., Fok, M. V. and Chukova, Yu. P., "Using Anti-Stokes Luminophors in Low-Power Light Sources", IZVESTIYA AN SSSR, SER. FIZ., Vol 40 pp 2372-2375, 1976

96. Kovalenko, V. F., Lisenker. B. S., Lisovenko, V. D. et al, "Gallium Arsenide Electroluminescent Diodes With Anti-Stokes Luminophors", FTP Vol 12 pp 258-263, 1978
    Antonov-Romanovskiy, A. V., Levshin, V. L., Morgenshtern, Z. L. and Trapeznikova, Z. A., "Investigation of Extremely IR-Sensitive Alkali-Earth Phosphors", ZhETF Vol 17 pp 949-956, 1947

97. Antonov-Romanovskiy, A. V., Levshin, V. L., Morgenshtern, Z. L. and Trapeznikova, Z. A., "Scintillation Mechanism in SrS Phosphors Activated With PZ Activators and the Interaction of the Activators", IZVESTIYA AN SSSR, SER. FIZ., Vol 13 pp 75-79, 1949

98. Arapova, E. Ya., Kut'yenkov, A. A., Mitrofanova, N. V. et al, "Spectral and Energy Properties of ZnS-Cu,Me Luminophors During UV and IR Irradiation" in the book "Materialy XXV soveshchaniya po lyumines-tsentsii" [Proceedings of the 25th Conference on Luminescence], Lvov, Isd-vo Lvov univ-ta, 1978, 148 pp

99. Babkina, T. V., Gayduk, M. I., Zolin, V. F. et al, "Spectroscopy of Sm^{3+} and Pr^{3+} in Oxysulfides and the Role of O^{2-} and S^{2-} Anions in the Crystal Fields of Rare-Earth Ions" in the book "Spektroskopiya kristallov" [Crystal Spectroscopy], Moscow, Nauka, 1975, pp 292-295

100. Timofeyev, Yu. P. and Fridman, S. A., "Luminescence Conversion from IR to UHF Radiation Into Visible Light and Its Applications", IZVESTIYA AN SSSR, SER. FIZ., Vol 43 pp 1303-1312, 1979

101. Zemskov, K. I., Isayev, A. A., Kazaryan, M. A. and Petrash, G. G., "Laser Projection Microscope", KVANTOVAYA ELEKTRONIKA Vol 3 pp 35-41, 1976

102. Arapova, E. Ya., Isayaev, A. A., Kazaryan, M. A. et al, "Infrared Projection Microscope", KVANTOVAYA ELEKTRONIKA Vol 2 pp 1568-1570, 1975

103. Zemskov, K. I., Kazaryan, M. A., Petrash, G. G. et al, "Laser Projection Microscope With Barium Vapor Amplifier and Luminescent Screen for Visualization Of IR Images", KVANTOVAYA ELEKTRONIKA Vol 7 pp 2454-2459, 1980

104. Bazhulin, A. P., Irisova, N. A., Sasorov, V. P. et al, "The Radiovizor — A Device for Visual Observation and Recording Fields of IR-to-UHF Radiation", VESTNIK AN SSSR Vol 12 pp 15-22, 1973

105. Levshin, V. L., Mitrofanova, N. V., Timofeyev, Yu. P. et al, "Application of Crystal Phosphors for Registering Electromagnetic Radiation", TRUDY FIAN SSSR Vol 59 pp 64-123, 1972

106. Levshin, V. L., Mitrofanova, N. V., Timofeyev, Yu. P. et al "Temperature Sensitive Luminophor for Visualization of Millimetric Radiation", PTE Vol 4 pp 166-167, 1970

107. Timofeyev, Yu. P. and Fok, M. V., "Numerical Calculation Of the Kinetics of Emission in Temperature-Sensitive Crystal Phosphors", TRUDY FIAN SSSR Vol 129 pp 143-171, 1981

108. Vinokurov, L. A. and Fok, M. V., "On the Quenching of ZnS:Cu,Co and ZnS:Cu,Ni Phosphors by Infrared Light", OPTIKA I SPEKTROSKOPIYA Vol 1 pp 248-254, 1956

Spectroscopy of ZnS-Tm
Luminescence Centers

N.N. Grigor'yev, M.V. Fok,

L. Yastrabik

Abstract: ZnS crystals were synthesized, whose room temperature luminescence spectrum consists of a series of bands 0.03—0.04 eV wide belonging to
Tm^{3+}, and broad structureless bands with a maximum in the blue region of
the spectrum caused by centers of other origin. At helium temperatures the
narrow bands split into a set of lines which differ in intensity (by up to
three orders of magnitude) and degree of polarization (exceeding 50% for some
lines), as well as their sensitivity to infrared radiation. Room temperature
investigations of band intensity relaxation after pumping or infrared radiation
were switched on and off, as well as investigations of the dependence of this
intensity on the pumping intensity, showed that Tm creates luminescence
centers in ZnS which are distinguished by their structure. It is shown that
these centers take on energy in the following ways: by resonance migration
from blue luminescence centers, by direct absorption of excitation photons,
by pure recombination, and through intermediate levels in two-stage
excitation. On the basis of analysis of the location of more than 200 bands
and the temperature dependence of many of these (at temperatures ranging
from that of helium to nitrogen), we constructed a system of sublevels for a
number of lower Tm terms corresponding to transitions between these lines.
We identified more than two-thirds of the lines in this investigation this
way. We indicate the course of future investigations needed for a complete
spectral identification of Tm in these crystals.

In investigating the luminescence mechanism of a new crystal phosphor,
one has to answer the same "initial questions": 1) what kind of centers does
the given activator form in the crystal? and 2) How do these centers recieve

pumping energy? For rare-earth and some other activator types we must add yet a third question to this: which electronic transitions correspond to one or another emission band? In this article, an attempt is made to answer these questions to some degree for the relatively unstudied ZnS—Tm crystal phosphor, which at room temperature has a series of narrow emission bands ranging from the indigo to the infrared parts of the spectrum.

Since all three of these questions are very closely tied to one another, we shall have to answer them, so to speak, by successive approximation — first we will answer all three questions in broad outline, and then in more detail, using the answer we have obtained for two of the other questions while trying to answering the third.

SAMPLES AND METHOD

We studied single crystals and, in selected cases, powdered ZnS—Tm luminophors. The crystals were grown by sublimation at 1320—1350° C from a deoxidized melt containing Tm in concentration of $10^{-2}-10^{-3}$ wt.%. The activator was doped into the melt in the form of $Tm(NO_3)_3$ salts. A quartz ampule with the melt in it was loosely sealed with a stopper in order to make diffusion exchange of gases with the environment more difficult yet not prevent removal of gases which are released (in particular, desorbed oxygen and water, as well as sulfur oxides, etc.) and thereby help maintain the pressure inside the ampule at close to atmospheric. The ampule was placed in a quartz tube, through which a slow stream of dessicating hydrogen sulfide was passed, then the tube was placed in a tube furnace and progressed through it over a period of 30—40 h (for more details cf. [1]). The crystals grew on the end of the ampule as it moved away from the hottest furnace regions. It turns out that the presence of oxygen strongly impedes Tm luminescence: the degree of deoxidation achieved by preliminary treatment of the melt in hydrogen sulfide was not enough to suppress the blue-green band characteristic of oxygen-containing zinc sulfide. In addition, some fraction of oxygen may penetrate to the crystal-growth atmosphere from the ampule walls or even, via diffusion, from the reaction-tube atmosphere through the heated walls. Against the background of this band the indigo luminescence of Tm is barely discernable (Fig. 1a) and may only be reliably measured at nitrogen or even lower temperatures. Thus an admixture of sulfur or carbon was doped into the melt in quantity of 5 wt.%, which promotes the bonding of oxygen. The vapor forms of these oxides are difficult to dissociate and are rather quickly removed from the ampule. We made no use of monovalent co-activators. All the crystals were quasi-cubic and contained from 15—20% hexagonal phase, which was determined by the "fluted" or "grooved" spectrum method of [1].

Luminescence was excited by means of: individual mercury lines (313, 365, 405, 436 nm) from a DRSh-250 or DRSh-100 lamp or a combination of these lines obtained from light filtering; a part of the radiation from a deuterium or 200 W xenon flashtube, selected out by an MDR-2 or UM-2 monochromator; or, finally, radiation from a tuneable dye laser (the "Molektron") with a pulse width of 8 ns. The luminescence spectra were measured on a low-resolution monochromator (UM-2) or on an SDL-1 double spectrometer with interchangeable gratings covering the 220—650 nm region with resolution on the order of 0.8—1.6 nm/mm. FEU-79 or FEU-62 photoelectron multipliers served as photodetectors in either case, and their signal was fed to a synchronous detection amplifier and recorder. For pulsed exitation, the spectrum was registered with the aid of a boxcar integrator.

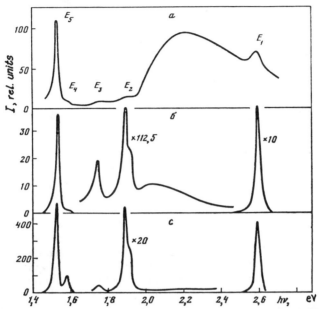

Fig. 1   Luminescence Spectra of ZnS—Tm Samples Obtained Under Different Preparation Conditions. Pumped by 365 nm (3.39 eV) Radiation at Room Temperature.   a- ZnS-Tm,O;   b- ZnS-Tm,S;   c- ZnS-Tm,C.

The investigated crystal was pumped over the (110) face and the luminescence from that face was recorded. The angle between the beam of pumping radiation and the beam of luminescence light was approximately 30° (in air). We also investigated powdered samples for reflection. We conducted measurements, as a rule, near room, nitrogen or helium temperatures. During the investigation of temperature dependence of the luminescence intensity of individual fine structure lines near the temperature of liquid helium, the entire section of the spectrum under study was recorded, allowing us to study temperature diplacement of line positions. During polarization measurements the sample was rotated about an axis perpendicular to the (110) cleavage plane, and the analyzer in front of the monochromator slit remained stationary. This allowed us to introduce corrections to the polarization characteristics of the monochromator and electron multiplier. In experiments on the effect of preliminary infrared irradiation we used the radiation of an incandescent lamp (500 W) in the interval 0.4 eV $< h\nu <$ 1.4 eV.

LUMINESCENCE CENTERS

The room temperature luminescence spectra of our samples had five narrow (halfwidths of 0.03—0.04 eV) bands with photon energy at maximum of $E_1 = 2.59$ eV, $E_2 = 1.90$ eV, $E_3 = 1.77$ eV, $E_4 = 1.59$ eV and $E_5 = 1.54$ eV which may be ascribed to Tm^{3+}, and the already-mentioned broad structureless band in the 1.9—2.8 eV region apparently belonging to oxygen-containing centers (see Fig. 1). Upon reduction of temperature the narrow bands split, changing into groups of lines. This is easy to see from Fig. 2, where part of the spectrum in the 1.61—1.73 eV region is shown, obtained at 77 K for an array of energies of pumping radiation photons. At temperatures lower than 77 K a new group of lines appears with E_6 close to $h\nu = 1.84$ eV.

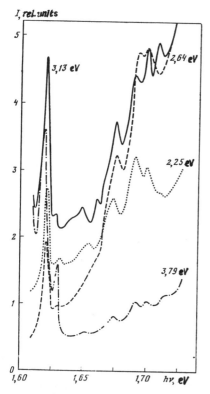

Further reduction in temperature lowers the relative intensity of the "oxygen" bands so that the lines show through even clearer. At the temperature of liquid helium the bands become very numerous. A small part of the spectrum is schematically depicted in Fig. 3 corresponding to the E_1 band, taken at 4.9 K. It contains around 60 lines. Dozens of lines are observed also in other parts of the spectrum, corresponding to the remaining bands. This shows that the observed bands $E_1 - E_6$ do indeed belong to a rare-earth element, in this case Tm.

Fig. 2. Part of the Photoluminescence Spectrum of ZnS—Tm,S at 77 K During Pumping by Photons of Different Energy

Numbers near the curves are the energy of excitation photons in eV

Fig. 3. Schematic Depiction of a Part of the Spectrum in the Region of the E_1 Band ($h\nu_{max} = 2.59$ eV) During Pumping by 330 nm Radiation (3.75 eV) at 4.9 K.

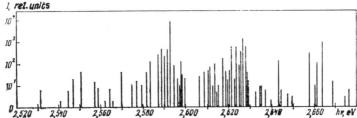

76

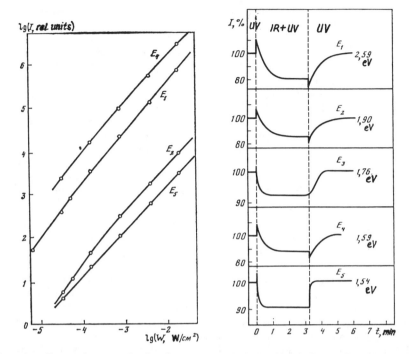

Fig. 4. Dependence of the Intensity of Individual ZnS—Tm,S Luminescence Bands on the Pumping Radiation Intensity with Photon Energies Around 3.75 eV at Room Temperature. Maxima of bands E_1=2.59 eV, E_2=1.90 eV, E_4=1.59 eV, E_5=1.54 eV

Fig. 5. Effect of Infrared Illumination (0.4 eV $<h\nu<$1.4 eV) on the Intensity of the E_1, E_2, E_3, E_4 and E_5 Bands. Dashed lines indicate the moment of switching on and off the IR illumination. Pumping is by photons with $h\nu$=3.95 and 3.40 eV at room temperature. ZnS-Tm,S sample.

     As a rule, an activator ion in Zns can form luminescence centers of several types, depending on its surroundings. (For example, by the relative location of ions of other impurities or base lattice defects.) All these centers may be differentiated by the symmetry or strength of the crystal field, which also depends on the distance the luminescence center is located from (or within) a hexagonal interlayer. On the other hand, this field defines the probabilities of transitions between sublevels of Tm^{3+} terms. Since the relative concentrations of various types of centers depends on the preparation conditions, the different ZnS—Tm samples can differ strongly in the relative line intensity of the luminescence spectrum.

     As was already mentioned, Fig. 1 represents typical luminescence spectra obtained on a low-spectral-resolution device at romm temperature for crystals grown under various preparation conditions at 1350° C. As was already noted above, the presence of excess sulfur (Fig. 1b) or carbon (Fig. 1c) in the crystal-growth atmosphere sharply lowers the intensity of the broad structureless band (Fig. 1a) which is linked to residual oxygen. However, even on the noisy background of this band, we discern maxima located coincident with the E_1-E_4 bands enumerated above. During the conversion to crystals of ZnS−Tm and ZnS−Tm,C these bands become even more distinct. Along with this, and independent of growth conditions, the relative intensity of the bands changes while its spectral location and halfwidth remain constant. Thus the E_1 and E_5 bands which attain approximately the same intensity in ZnS−Tm,C differ by about a factor of 10 for ZnS−Tm,S and still more in ZnS−Tm,O. At the same time, the E_2 and E_3 bands in ZnS−Tm,S are much closer in intensity (a difference of about a factor of 2) than they are in ZnS−Tm,C, where they differ by approximately a factor of 8. The E_4 is more distinctly expressed in ZnS−Tm,C than in ZnS−Tm,S. We remark in passing that the presence of carbon in the growth atmosphere leads to much stronger overall emission of the crystal than does an excess of sulfur, which points to a difference in relative concentrations of quenching centers too, not just luminescence centers.

The existence of various sorts of luminescence centers in ZnS−Tm is also revealed in the luminescence kinetics of the individual bands. Thus their intensities, generally speaking, will depend in different ways on the pumping intensity (Fig. 4) and will relax at different rates after it changes. But most revealing in this regard is the effect of infrared radiation on the intensity of individual bands. After it is turned on, scintillation is observed in all bands E_1-E_5 with subsequent quenching of the emission (Fig. 5). There, however, the similarity ends in the behavior of the individual bands. As is apparent from the figure, relaxation of brightness proceeds differently in different spectral bands. After the switching off of the infrared radiation, the E_1, E_2 and E_4 bands have brightness minima that the E_3 and E_5 bands do not. Thus we may think that the fundamental contribution in this is made by the different structures of each of the centers. However, the behavior of these same E_3 and E_5 bands differs strongly: in the E_5 band, the brightness is restored after switching off the infrared radiation much more quickly than in E_3. This indicates that they too, apparently, belong to different centers.

ENERGY TRANSFER TO LUMINESCENCE CENTERS

In photoluminescence, generally speaking, there are several possible routes by which a Tm ion can recieve the energy leading to its excitation: direct absorption by the Tm^{3+} ion of an excitation photon − Route I ("intra-center"); a pure recombination route, wherein nonequilibrium charge carriers arising during pumping recombine directly at the centers formed by Tm, and transfer to them the energy thus given off − this is Route II; and the sensitization route, wherein a center-sensitizer is first excited (directly or due to recombination), which then transfers energy to a Tm ion by induced resonance − Route III. These routes are not mutually exclusive, and the relationship between them may depend on excitation conditions and the conditions of ZnS−Tm crystal synthesis.

In order to clarify whether the Tm emission is recombination or intra-center, we must use the results of the experiments, in part already described

78

above. Strictly speaking, this is exactly why they were done, for a clarification on this question. In particular, towards this end we investigated the dependence of luminescence intensity of a powdered luminophor (ZnS—Tm) on the pumping intensity for the Tm bands which were most intense at room temperature. Excitation was carried out by photons with energy of 3.75 eV. It turns out that this dependence is highly linear and slightly different for each emission band (Fig. 4). This dependence is the same in the E_1, E_2 and E_4 bands and is described by a step function with a value of 1.3—1.4 for pumping densities up to $3 \cdot 10^{-4}$ W/cm^2, then decreasing down to 1.06—1.07 at pumping densities higher than $3 \cdot 10^{-3}$ W/cm^2, which is readily visible from the discontinuity in the log-log curve describing this function. The E_5 band is distinguished by the fact that this function is constant for all investigated ranges of pumping density and is equal to approximately 1.04.

Not excluding the possibility that intra-center excitation mechanisms are manifest in several cases, in particular for the E_5 band, this linearity unequivocally indicates the substantial role of recombination mechanisms in the luminescence. This is also indicated by the long (lying in the range of seconds , rather than nano-, micro- or even milliseconds) relaxation of

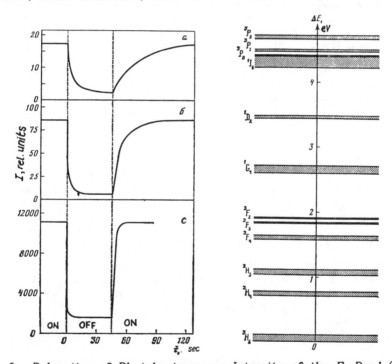

Fig. 6. Relaxation of Photoluminescence Intensity of the E_1 Band ($h\nu = 2.59$ eV) for Various Intensities of Pumping Radiation of $h\nu = 3.07$ eV at Room Temperature. The Sample is ZnS—Tm,S. The dashed lines indicate the moment the pumping radiation is switched on and off: a- Intensity $2.9 \cdot 10^{-7}$ W/cm^2; b- $1.3 \cdot 10^{-6}$ W/cm^2; c- $5.3 \cdot 10^{-5}$ W/cm^2

Fig. 7. Term Diagram of Tm^{3+} in Y$_2$O$_3$ According to the Data of [2]
The width of the strips corresponds to term splitting

79

luminescence intensity of powdered ZnS—Tm in each of the bands after a sharp reduction in pumping density by three orders of magnitude, as well as an acceleration in relaxation with the growth of pumping intensity, after the pumping radiation is switched on or off, as shown in Fig. 6 for the E_1 band.

For further study of the centers we need to know, even if only in broad outline, the term system of Tm^{3+}. The positions of the centers of gravity of Tm^{3+} terms and their splitting in Y_2O_3—Tm crystals are given in [2]. However, transferring this data over to ZnS—Tm should be done with caution, since the strength and symmetry of that crystal field may be quite different. This above all would be expressed in the ratios of transition probabilities between sublevels, as well as in the splitting of terms. Thus we will not take these into account, and employ only the positions of the centers of gravity, which are less sensitive to the crystal field.

As can be seen from Fig. 7, there are a few term pairs that have little space between them. This leads to ambiguity in assigning the experimentally-observed bands to the determined electronic transition. Thus the E_5 band with $h\nu_{max}=1.54$ eV ascribed in [3] to the $^1G_4\rightarrow^3H_5$ transition has a possible energy in the 1.44—1.56 eV region, in agreement with the proposed term scheme, while in [4] it is ascribed to the $^3F_2\rightarrow^3H_6$ with a possible energy of 1.42—1.60 eV. However, it may be assigned with equal success to the $^1D_2\rightarrow^3F_2$ transition for which the spacing between terms may oscillate within the range of 1.45—1.50 eV. The E_6 band with $h\nu_{max}=1.84$ eV is ascribed in [4,5] to the $^3F_2\rightarrow^3H_6$ transition (term spacing 1.84—1.96 eV), but according to [6] it may be the $^1D_2\rightarrow^3F_4$ transition (1.80—1.85 eV). The E_1 band with $h\nu_{max}=2.59$ eV may be assigned to both the $^1G_4\rightarrow^3H_4$ transition (2.52—2.72 eV) and to the $^1I_6\rightarrow^3F_4$ transition (2.56—2.84 eV). Similar arguments can be made as regards the other transitions.

We will now examine possible variants in excitation of the Tm^{3+} term series. The exact location of the lower term $^3H_6$ of the Tm^{3+} ion in zinc sulfide relative to the valence band is unknown. However, based on research conducted in [7] into the kinetics of the polarization of luminescence from ZnS—Tm crystals in the E_2 band in connection with transitions based on the $^3H_6$ term, we can consider that the recombination of holes with electrons localized on Tm^{3+} ions does in fact take place. Therefore one can state that the $^3H_6$ term is located quite close to the ZnS valence band.

The upper levels $^3P_0$, $^3P_1$, $^3P_2$, and $^1I_6$ are located 4.2 eV higher than the ground term $^3H_6$, i.e. by a greater spacing than the width of the forbidden band, which in ZnS is approximately equal to 3.7 eV. Therefore both recombination mechanisms for excitation of these levels are excluded. The first of these methods is inhibited because in this region the absorption of the zinc sulfide itself is great and the pumping radiation will hardly make it all the way to the Tm ions. Nonetheless, we can detect the excitation of such levels if we compare the luminescence spectra in the indigo region for pumping radiation with photons of various energies. In Fig. 8 it is apparent that for excitation by photons of 4.67 and 4.28 eV, corresponding to the $^3H_6\rightarrow^3P_3$ and $^3H_6\rightarrow^1I_6$ transitions, the luminescence spectrum has an intensive line at 2.592 eV, two weak ones at 2.611 and 2.629 eV and a "shoulder" around 2.593—2.595 eV, that are all absent during excitation by photons of $h\nu=3.68$ eV, which are unable to excite terms lying above $^1D_2$. Since all these lines appear during excitation by photons with energies of both 4.67 and 4.28 eV, then they should be ascribed to transitions from the $^1I_6$ level, i.e. $^1I_6 \rightarrow {}^3F_4$. This group of lines and its corresponding bands at room temperature we will call E_7, (We did not succeed in observing lines corresponding to transitions from any sort of ^3P state).

The next term $^1D_2$ is located, according to [2], approximately 3.45 eV above the lower sublevel of the ground term $^3H_6$. According to our identification (which we discussed earlier), it is located roughly 0.2 eV higher than that, but even so this spacing does not exceed the width of the forbidden band in zinc sulfide. Therefore if the lower term is located close enough to the valence band, then excitation of the $^1D_2$ term is possible by the recombination Route II, i.e. by recombination of holes with electrons localized on Tm^{3+} ions. As we will see further on, this is just what happens in a number of cases. A term is most easily excited at photon energies close in value to the width of the forbidden band.

The $^1G_4$ term, located 2.6 eV above the ground level, may be excited by both Route II, i.e. recombination, and by Route III, i.e. intercepting the energy from a blue luminescence center (which has a spectrum covering this region) by means of a direct inductive resonance mechanism, and taking on this energy in turn by recombination. In the excitation spectrum for $E_1 - E_5$ band luminescence there is indeed a band with a maximum of 3.75 eV and a half-height width of around 0.25 eV (Fig. 9), which is characteristic of the blue luminescence centers in zinc sulfide. As is apparent from this figure, the excitation spectra of the E_1, E_2 and E_4 bands are very similar to one another and differ markedly from the excitation spectra of the E_3 and E_5 bands which are likewise similar to each other, although to a lesser extent.

That the band with a maximum at 3.75 eV in the adduced excitation spectra for the E_1, E_2 and E_4 bands is in fact linked to the excitation of the blue luminescence centers in zinc sulfide is also confirmed by the temperature dependence of the excitation spectra. Thus for the E_2 band during the transition from helium to nitrogen temperatures, the relative height of the band with $h\nu_{max} = 3.75$ eV in the excitation spectrum hardly changes at all, and for further increase up to room temperature it decreases by several hundred percent. As is known, this behavior is similar to that in the blue luminescence band of "self-activated" ZnS, which experiences quenching at room temperature.

The E_1, E_2 and E_4 bands are close to one another also in terms of their kinetic properties. As already noted, they are united by their dependence on pumping intensity (Fig. 4). In addition, for higher than room temperatures and excitation in the 3.75 eV band they are all quenched with activation energy $\Delta E = 0.58$ eV and frequency factor $A = 2 \cdot 10^{10}$. The E_5 band is characteriazed by $\Delta E = 0.21$ eV and $A = 3.5 \cdot 10^3$. In terms of the intensity relaxation rate (in the range of seconds) after a sharp decrease (three orders of magnitude) in the pumping density with 3.75 eV photons, the E_1, E_2 and E_4 bands were again different from the E_5 band (Fig. 10a). All this would make us think that they share a common excitation mechanism and even that they result from a common upper term. On this basis, and taking the adduced term scheme into consideration, the E_1 band ($h\nu_{max} = 2.59$ eV) may be ascribed to the $^1G_4 \rightarrow {}^3H_6$ transition, the E_2 band ($h\nu_{max} = 1.90$ eV) to the $^1G_4 \rightarrow {}^3H_4$ transition, and the E_4 band ($h\nu_{max} = 1.59$ eV) to the $^1G_4 \rightarrow {}^3H_5$ transition.

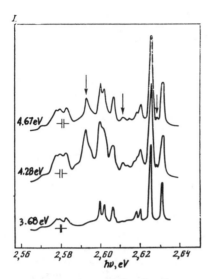

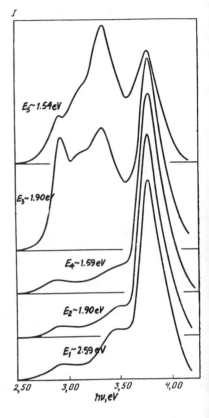

Fig. 8. Part of the ZnS−Tm,S Spectrum During Pumping With Photons of Various Energies (as Indicated Near the Curves) at 77 K

The arrows indicate lines generated by excitation photons with energy greater than 3.68 eV.

Fig. 9. Excitation Spectra of Emission Bands E_1 Through E_5 With Maxima Indicated Next to Each Curve

The sample is ZnS-Tm,S at 77 K

Fig. 10. Relaxation of the Intensity of the E_1, E_2, E_3 and E_5 Bands During Pumping by Photons of Various Energies After a Decrease ($\times 10^{-3}$) in Intensity of Pumping Radiation.

a- From 10^{-1} to $1.5 \cdot 10^{-4}$ W/cm², $\hbar\nu$=3.75 eV;
b- Decrease from $4 \cdot 10^{-3}$ to $5 \cdot 10^{-6}$ W/cm²,
$\hbar\nu$=3.06 eV

Sample is ZnS-Tm,S at room temperature

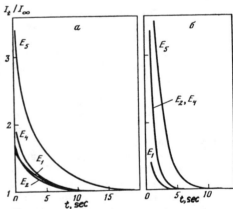

During excitation by an aggregate of photons having several different energies the similarity in behavior of the E_1, E_2 and E_4 bands breaks down, and their thermal quenching deviates sharply from the Mott formula. (For example, during excitation by photons of 3.40, 3.06 and 2.84 eV the E_5 band, together with those in question, has a minimum in its intensity as a function of temperature.) Apparently, it happens that lines belonging to other transi-

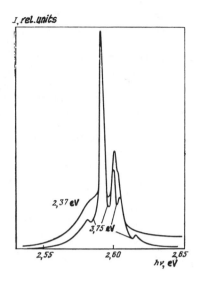

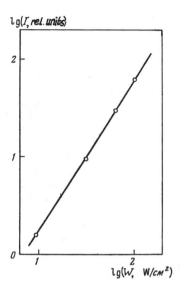

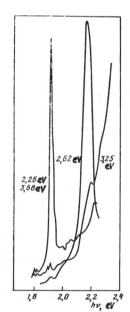

Fig. 11. Luminescence Spectrum of ZnS—Tm in the Region of $h\nu \approx 2.6$ eV During Stokes (3.75 eV) and Anti-Stokes (2.37 eV) Pumping at 100 K

The photon energy of the pumping radiation is indicated next to the curve

Fig. 12. Dependence of the Luminescence Intensity in the E_1 Band ($h\nu_{max}=2.59$ eV) on the Intensity of Anti-Stokes Pumping with $h\nu=2.37$ eV

The sample is ZnS-Tm,S at 100 K

Fig. 13. Comparison of the Luminescence Spectra in the Long-Wavelength Part of the Visible Spectrum for the Case of Pumping With Photons of Various Energies (Indicated Next to the Curve) at Room Temperature

tions which may be excited by some other mechanism fall in this same region of the spectrum. The observed (Fig. 10b) differences in the relaxation rates of the E_1 band compared to E_2 and E_4 for excitation photons with energy $h\nu=3.06$ eV also attest to this breakdown.

As demonstrated in our experiments Route I, i.e. intra-center excitation, apparently comes in two varieties — besides the excitation by one photon of sufficiently high energy, excitation is also possible by two photons of lesser energy through intermediate levels. For example, Tm luminescence at $h\nu=2.59$ eV is also excited by photons of 2.48, 2.36 and even 2.27 eV. The energy difference is so great that even at room temperature it could make up for thermal motion, and this luminescence is observed as well at 100 K (Fig.

11). Fig. 12 shows the intensity of $h\nu = 2.59$ eV luminescence as a function of anti-Stokes pumping intensity on log-log axes. At 100 K it is described well by a step function with a value of 1.67, while at room temperature this value reaches 2.3. During Stokes pumping, as has already been noted, it is about 1.3. This difference confirms the idea that some sort of intermediate level is active during anti-Stokes pumping. It is located, apparently, somewhere just a little less than 2.3 eV above the ground level of the luminescence center. The behavior of the dependence of the luminescence spectrum on the energy of excitation phonons also indicates the presence of such a level. As is apparent from Fig. 13, the narrow E_2 band (1.9 eV), which according to the scheme in Fig. 7 may be ascribed to the $^1D_2 \rightarrow ^3F_3$, $^1G_4 \rightarrow ^3H_4$ or $^3F_2 \rightarrow ^3H_6$ transitions, is excited not only by photons of $h\nu = 3.68$ eV corresponding to the $^3H_6 \rightarrow ^1D_2$ transition but also by 2.26 eV photons, whereas 2.62 eV photons corresponding to the $^3H_6 \rightarrow ^1G_4$ transition and 3.25 eV photons, which are not resonant to any Tm^{3+} transitions, hardly cause any excitation, although they do excite luminescence in the $2.16 - 2.19$ eV range. Apparently the $^3F_2$ term may receive pumping energy during the transition from the 2.3 eV level, and the $^1G_4$ term, conversely, gives its energy to it. The nature of this level is, however, unknown in the meantime, since it is not a part of the Tm^{3+} term system that is in acceptance today.

LOW-TEMPERATURE SPECTRA

At room temperature and even at liquid nitrogen temperature, the lines corresponding to transitions between individual sublevels of Tm^{3+} terms are so broad that they run together and merge into bands that yield to analysis only with difficulty. Thus the investigation was undertaken at much lower temperatures. We investigated the emission spectra at 4.9 K, the degree of spontaneous luminescence polarization at 6.5 K and the effect of infrared radiation at 8.2 K. The predominant direction of the E-vector of the luminescence radiation was always either parallel or perpendicular to the optical axis of the crystal (the C axis). In the former case the polarization was defined as positive, and in the latter, as negative. The effect of infrared radiation (0.4 eV $< h\nu < 1.4$ eV) was characterized by a quenching coefficient K equal to

$$K = \frac{I(0) - I(IR)}{I(0)} \quad ,$$

where $I(IR)$ and $I(0)$ are the line intensities in the presence of infrared radiation and without it. In this way $K > 0$ signifies infrared-stimulated quenching and $K < 0$ signifies a build-up. The results obtained are reduced in Tables $1 - 3$. In the last column of the Tables we indicate the transitions between terms which we were able to assign to fine structure lines, based on the identifications presented in this article and described above.

Glancing at these Tables, the first thing to catch your eye is the abundance of lines. In total, we managed to measure the location of 228 lines. If one were to consider that only those transitions from a little more than the lowest sublevels are possible, then there would be at least half as many still. On the other hand, two hundred lines in the interval from 1.5 to 2.7 eV is not so many. This quantity of lines does not however prove that they belong to centers of several different sorts. Actually, the probability of

Table 1. Radiation Spectrum of ZnS—Tm in the Indigo Region at 4.9 K

Wavelength λ in Å	Photon Energy $h\nu$ in eV	Intensity in arbitrary units	Quenching Coefficient[1] K, in %	Polarization[2] P in %	Transitions
4632,8	2,6762	6			
4635,9	2,6744	3			
4646,1	2,6686	14			$^1G_4 \to {}^3H_6$
4652,4	2,6650	—*			$^2G_4 \to {}^3H_6$
4653,8	2,6642	940	—17,5	—65	
4654,2	2,6611	100	0	—28	
4660,8	2,6602	10			$^1G_4 \to {}^3H_6$
4665,3	2,6576	270	+20	—54	
4679,9	2,6493	3			
4683,2	2,6474	4			$^1G_4 \to {}^3H_6$
4688,7	2,6443	6			
4690,5	2,6433	135	0	—56	
4698,5	2,6388	2			
4701,9	2,6369	6			$^1G_4 \to {}^3H_6$
4705,4	2,6350	9			$^1G_4 \to {}^3H_6$
4706,6	2,6343	9		—88	$^1G_4 \to {}^3H_6$
4709,6	2,6326	5			$^1G_4 \to {}^3H_6$
4712,0	2,6312	—*			$^1G_4 \to {}^3H_6$ $^1D_2 \to {}^3H_5$
4716,3	2,6289	15			$^1G_4 \to {}^3H_6$
4716,9	2,6285	50		—54	
4717,7	2,6281	650	—46	—58	
4720,6	2,6265	1450	—24	0	$^1G_4 \to {}^3H_6$ $^1D_2 \to {}^3H_5$
4722,2	2,6256	380	+3	—13	
4723,3	2,6250	90	0	+90	$^1G_4 \to {}^3H_6$
4726,0	2,6235	600	+6	+8	$^1G_4 \to {}^3H_6$
4727,2	2,6228	20		+10	
4728,0	2,6204			—45	$^1G_4 \to {}^3H_6$
4729,0	2,6218			—52	$^1D_2 \to {}^3H_5$
4730,7	2,6208	640	0	—52	
4731,5	2,6204	50		—48	$^1G_4 \to {}^3H_6$
4732,2	2,6200				
4733,2	2,6195	20			$^1G_4 \to {}^3H_6$
4734,5	2,6188	50	0	0	$^1G_4 \to {}^3H_6$
4735,8	2,6180				$^1D_2 \to {}^3H_5$
4737,4	2,6172	160	+21	+20	$^1G_4 \to {}^3H_6$
4741,1	2,6151	20			$^1G_4 \to {}^3H_6$ $^1D_2 \to {}^3H_5$
4742,3	2,6144	5			$^1G_4 \to {}^3H_6$
4744,0	2,6135	100	+22	—49	$^1G_4 \to {}^3H_6$
4745,6	2,6126	10			$^1G_4 \to {}^3H_6$
4748,2	2,6112				$^1G_4 \to {}^3H_6$
4748,6	2,6110	70	—24	—90	$^1G_4 \to {}^3H_6$ $^1D_2 \to {}^3H_5$
4749,7	2,6104	50	—25	—90	$^1G_4 \to {}^3H_6$ $^1D_2 \to {}^3H_5$
4753,4	2,6083	40	—22	+20	$^1G_4 \to {}^3H_6$
4756,8	2,6066				$^1G_4 \to {}^3H_6$
4757,5	2,6061	25			$^1G_4 \to {}^3H_6$ $^1D_2 \to {}^3H_5$
4766,0	2,6014				$^1G_4 \to {}^3H_6$ $^1D_2 \to {}^3H_5$
4769,3	2,5996	20			$^1G_4 \to {}^3H_6$ $^1D_2 \to {}^3H_5$
4772,7	2,5978	30			$^1G_4 \to {}^3H_6$
4773,4	2,5974	120	0	0	$^1G_4 \to {}^3H_6$ $^1D_2 \to {}^3H_5$
4774,3	2,5969	10			$^1D_2 \to {}^3H_5$

Table 1. (Continued)

Wavelength λ in Å	Photon Energy $h\nu$ in eV	Intensity in arbitrary units	Quenching Coefficient[1] K, in %	Polarization[2] P in %	Transitions
4775,8	2,5961	20	0		$^1G_4 \to {}^3H_6$
4779,1	2,5943	20	0	+8	$^1G_4 \to {}^3H_6$ $^1D_2 \to {}^3H_5$
4781,9	2,5928				$^1G_4 \to {}^3H_6$
4782,6	2,5924	7840	−23	−38	$^1G_4 \to {}^3H_6$
4784,4	2,5914	450	0	−55	$^1D_2 \to {}^3H_5$
4786,8	2,5901	230	−20	+65	$^1G_4 \to {}^3H_6$
4789,3	2,5888	490	−25	0	
4790,0	2,5884			0	$^1G_4 \to {}^3H_6$
4792,5	2,5871	260	+11	+61	$^1G_4 \to {}^3H_6$
4799,7	2,5832	120	−14	+15	
4801,0	2,5825				$_1G_4 \to {}^3H_6$
4802,7	2,5816	40		+31	$^1G_4 \to {}^3H_6$
4806,9	2,5793	10			$^1G_4 \to {}^3H_6$
4811,7	2,5767	15	0	0	$^1G_4 \to {}^3H_6$
4815,6	2,5746	10			$^1G_4 \to {}^3H_6$
4824,9	2,5697	40	+23	−58	$^1G_4 \to {}^3H_6$
4832,8	2,5655	2			$^1G_4 \to {}^3H_6$
4835,4	2,5641	7			$^1G_4 \to {}^1H_6$
4839,6	2,5619	2			$^1G_4 \to {}^3H_6$
4845,6	2,5587	8			
4848,6	2,5571	15			
4861,2	2,5505	44	−17	+13	
4863,0	2,5496				$^1G_4 \to {}^3H_6$
4868,0	2,5469	20			
4872,4	2,5446	6			
4879,4	2,5405	2			
4898,0	2,5313	6			
4941,6	2,5090				

1 Plus sign corresponds to quenching of the given intensity of infrared radiation, and a minus sign corresponds to an increase or build-up.

2 Plus corresponds to a predominant direction of the radiation E-vector parallel to the optical axis

• The position of these lines was determined at 15 K since they flare up and are not visible still at 4.9 K. Unfilled entries correspond to a lack of data.

Table 2. Radiation Spectrum of ZnS—Tm in the Red Region at 4.9 K

Wavelength λ in A	Photon Energy in eV	Intensity in arbitrary units	Quenching Coefficient[1] K in %	Polarization[2] P in %	Transition
6478,9	1,9137	580	−10	0	$^3F_2 \to {}^3H_6$ $^1G_4 \to {}^3H_4$ $^1D_2 \to {}^3F_3$
6483,3	1,9123	2			
6487,4	1,9112	360	−10	+57	$^1D_2 \to {}^3F_3$
6492,1	1,9098	1100	−12	+61	$^3F_2 \to {}^3H_6$ $^1G_4 \to {}^3H_6$ $^1D_2 \to {}^3F_3$
6496,8	1,9084	6			
6502,3	1,9068	140	−21	+47	$^3F_2 \to {}^3H_6$ $^1G_4 \to {}^3H_4$ $^1D_2 \to {}^3F_3$
6504,5	1,9061				$^3F_2 \to {}^3H_6$
6507,4	1,9053	340	−8	+60	$^1G_4 \to {}^3H_4$
6510,7	1,9043	110	−14		$^1G_4 \to {}^3H_4$ $^1D_2 \to {}^3F_3$

86

Table 2. (Continued)

Wave-length λ in A	Photon Energy in eV	Intensity in arbitrary units	Quenching Coefficient[1] K in %	Polar- ization[2] P in %	Transition
6515,8	1,9028	48	—14	+72	$^1G_4 \rightarrow {}^3H_4$ $^1D_2 \rightarrow {}^3F_3$
6522,4	1,9009	29	—9	—9	$_3F_2 \rightarrow {}^3H_6$ $^1D_2 \rightarrow {}^3F_3$
6527,2	1,8995	37	+14	+52	^{3} $^1G_4 \rightarrow {}^3H_4$ $^1D_2 \rightarrow {}^3F_3$
6531,6	1,8982	11	—20		$F_2 \rightarrow {}^3H_6$ $^1G_4 \rightarrow {}^3H_4$ $^1D_2 \rightarrow {}^3F_3$
6539,7	1,8959	45	—10	+21	$^3F_2 \rightarrow {}^3H_6$ $^1G_4 \rightarrow {}^3H_4$ $^1D_2 \rightarrow {}^3F_3$
6546,3	1,8940	97	—18	+49	$^3F_2 \rightarrow {}^3H_6$
6551,1	1,8926	23	0		$^3F_2 \rightarrow {}^3H_6$ $^1G_4 \rightarrow {}^3H_4$ $^1D_2 \rightarrow {}^3F_3$
6564,7	1,8887	15	—20	+4	$^1G_4 \rightarrow {}^3H_4$
6564,5	1,8873	5			$^3F_2 \rightarrow {}^3H_6$
6576,3	1,8853	180	—19	0	$^3F_2 \rightarrow {}^3H_6$ $^1G_4 \rightarrow {}^3H_4$
6586,2	1,8825	170	—11	0	$^1G_4 \rightarrow {}^3H_4$
6592,0	1,8808	69	0	+59	$^3F_2 \rightarrow {}^3H_6$
6603,6	1,8775	20	0	—24	$^3F_2 \rightarrow {}^3H_6$ $^1G_4 \rightarrow {}^3H_4$
6608,8	1,8761	71	+17		$^3F_2 \rightarrow {}^3H_6$
6614,7	1,8744	26	+21	—52	$^3F_2 \rightarrow {}^3H_6$
6618,7	1,8732	2			$^3F_2 \rightarrow {}^3H_6$
6638,0	1,8678	5			$^3F_2 \rightarrow {}^3H_6$ $^1G_4 \rightarrow {}^3H_4$
6648,7	1,8648	2			$^3F_2 \rightarrow {}^3H_6$ $^1G_4 \rightarrow {}^3H_4$
6718,2	1,8455	36	+16		
6722,8	1,8444	5			
6781,6	1,8282	14	+20		
6790,9	1,8258	3			
6843,7	1,8117	48	—70	—42	
6861,2	1,8070	3			
6868,3	1,8052	7			
6890,0	1,7995	9			
6908,0	1,7948	12			
6946,6	1,7848	230	—48	+52	$^3F_3 \rightarrow {}^3H_6$
6955,0	1,7828	600	+5	—24	$^1D_2 \rightarrow {}^3F_2$
6960,6	1,7812	20			$^3F_3 \rightarrow {}^3H_6$
6970,8	1,7786	2			$^3F_3 \rightarrow {}^3H_6$ $^1D_2 \rightarrow {}^3F_2$
6976,3	1,7772	80	—18		
6986,1	1,7747	1600	—20	—27	$^1D_2 \rightarrow {}^3F_2$
6999,0	1,7716	5			
7009,5	1,7688	220	+17	—20	$^1D_2 \rightarrow {}^3F_2$
7012,5	1,7680	90	0		$^1D_2 \rightarrow {}^3F_2$
7018,2	1,7662	2			$^1D_2 \rightarrow {}^3F_2$
7024,5	1,7650	230	+11	+7	
7032,5	1,7630	810	+20	—25	$^1D_2 \rightarrow {}^3F_2$
7034,5	1,7625		+17		
7037,3	1,7618	10			
7043,9	1,7608	60	+20		
7048,8	1,7590	90	—15	+47	$^1D_2 \rightarrow {}^3F_2$
7053,0	1,7579	30			$^1D_2 \rightarrow {}^3F_2$
7058,3	1,7566	35			$^3F_3 \rightarrow {}^3H_6$
7064,3	1,7551	130	—11		
7069,9	1,7537	10			$^3F_3 \rightarrow {}^3H_6$ $^1D_2 \rightarrow {}^3F_2$
7081,1	1,7509	10			$^3F_3 \rightarrow {}^3H_6$ $^1D_2 \rightarrow {}^3F_2$
7092,4	1,7481	4			$^3F_3 \rightarrow {}^3H_6$ $^1D_2 \rightarrow {}^3F_2$
7102,0	1,7458	3			$^3F_3 \rightarrow {}^3H_6$ $^1D_2 \rightarrow {}^3F_2$
7111,8	1,7434	7			$^3F_3 \rightarrow {}^3H_6$ $^1D_2 \rightarrow {}^3F_2$
7120,8	1,7412	4			$^1D_2 \rightarrow {}^3F_2$

Table 2. (Continued)

Wave-length λ in A	Photon Energy in eV	Intensity in arbitrary units	Quenching Coefficient[1] K in %	Polarization[2] P in %	Transition
7129,2	1,7391	3			$^3F_3 \rightarrow {}^3H_6$
7139,7	1,7366	6			$^3F_3 \rightarrow {}^3H_6$
7146,4	1,7349	3			$^3F_3 \rightarrow {}^3H_6$
7150,8	1,7339	9			$^3F_3 \rightarrow {}^3H_6$
7162,5	1,7310	6			$^3F_3 \rightarrow {}^3H_6$
7170,9	1,7290	8			$^3F_3 \rightarrow {}^3H_6$
7184,7	1,7257	2			$^3F_3 \rightarrow {}^3H_6$
7188,0	1,7249	3			
7191,2	1,7241,	4			$^3F_3 \rightarrow {}^3H_6$
7235,2	1,7136	7			

1 Plus sign corresponds to quenching of the given intensity of infrared radiation, and a minus sign corresponds to an increase or build-up.

2 Plus corresponds to a predominant direction of the radiation E-vector parallel to the optical axis of the crystal

Table 3. Radiation Spectrum of ZnS—Tm in the Near-Infrared at 4.9 K

Wavelength λ in Å	Photon Energy $h\nu$ in eV	Intensity in arbitrary units	Quenching Coefficient[1] K, in %	Polarization[2] P in %	Transitions
7652,7	1,6025	4*			
7694,8	1,6113	111	0		
7699,3	1,6103	4			
7703,0	1,6096	300	0	−10	$^3F_4 \rightarrow {}^3H_6$
7707,8	1,6086	66			
7714,4	1,6072	54			$^3F_4 \rightarrow {}^3H_6$ $^1G_4 \rightarrow {}^3H_5$
7720,7	1,6059	70			
7728,3	1,6043	62			
7735,6	1,6028	17			$^3F_4 \rightarrow {}^3H_6$ $^1G_4 \rightarrow {}^3H_5$
7742,0	1,6015	52			$^3F_4 \rightarrow {}^3H_6$
7755,2	1,5987	62			$^3F_4 \rightarrow {}^3H_6$
7767,0	1,5963	200	0	−9	
7770,0	1,5957				$^3F_4 \rightarrow {}^3H_6$
7772,5	1,5952				
7781,7	1,5933	42		+26	$^3F_4 \rightarrow {}^3H_6$
7790,9	1,5914	310	0	−25	$^1G_4 \rightarrow {}^3H_5$
7815,0	1,5865	2			$^1G_4 \rightarrow {}^3H_5$
7818,9	1,5857	10			$^3F_4 \rightarrow {}^3H_6$ $^1G_4 \rightarrow {}^3H_5$
7831,7	1,5831	45			$^1G_4 \rightarrow {}^3H_5$
7836,9	1,5821	94			$^3F_4 \rightarrow {}^3H_6$ $^1G_4 \rightarrow {}^3H_5$
7842,8	1,5809	12			
7847,0	1,5800	49			
7855,4	1,5783	250	0	+15	$^3F_4 \rightarrow {}^3H_6$ $^1G_4 \rightarrow {}^3H_5$
7859,0	1,5776	2800	+11		$^3F_4 \rightarrow {}^3H_6$
7867,1	1,5760	2350	+9	−10	$^3F_4 \rightarrow {}^3H_6$

Table 3. (Continued)

Wavelength λ in Å	Photon Energy hν in eV	Intensity in arbitrary units	Quenching Coefficient[1] K, in %	Polarization[2] P in %	Transitions
7875,7	1,5743	40			$^3F_4 \to {}^3H_6$ $^1G_4 \to {}^3H_5$
7883,0	1,5728	130	+14	−33	$^3F_4 \to {}^3H_6$ $^1G_4 \to {}^3H_5$
7894,6	1,5705	280	+12	−43	$^3F_4 \to {}^3H_6$ $^1G_4 \to {}^3H_5$
7908,9	1,5677				$^3F_4 \to {}^3H_6$ $^1G_4 \to {}^3H_5$
7912,0	1,5670	40			
7916,8	1,5661	60			$^3F_4 \to {}^3H_6$
7920,6	1,5654	60	0		$^3F_4 \to {}^3H_6$ $^1G_4 \to {}^3H_5$
7924,6	1,5646	160	0		$^3F_4 \to {}^3H_6$ $^1G_4 \to {}^3H_5$
7933,7	1,5628	80	+47		$^1G_4 \to {}^3H_5$
7937,4	1,5620	172	+23	+33	$^3F_4 \to {}^3H_6$ $^1G_4 \to {}^3H_5$
7941,4	1,5612	670	+33	+12	$^3F_4 \to {}^3H_6$ $^1G_4 \to {}^3H_5$
7947,0	1,5601	6 608	+31	−7	$^3F_4 \to {}^3H_6$
7948,8	1,5598	5 430		−31	$^3F_4 \to {}^3H_6$
7952,4	1,5591	150			$^3F_4^* \to {}^3H_6$ $^1G_4 \to {}^3H_5$
7954,7	1,5586	160			$^3F_4 \to {}^3H_6$
7957,7	1,5580	1 180			
7959,3	1,5577	1 020			$^3F_4 \to {}^3H_6$
7960,7	1,5575	1 160			$^3F_4 \to {}^3H_6$
7963,8	1,5568	1 400	+35	−36	$^3F_4 \to {}^3H_6$
7966,7	1,5563	196	0	0	
7970,8	1,5555	56			
7974,9	1,5547				$^3F_4 \to {}^3H_6$
7978,7	1,5540				$^3F_4 \to {}^3H_6$ $^1G_4 \to {}^3H_5$
7988,9	1,5520	150	0		$^3F_4 \to {}^3H_6$
7994,6	1,5509	86			$^3F_4 \to {}^3H_6$
7997,8	1,5502	20			$^3F_4 \to {}^3H_6$
8002,2	1,5494	20	+15	−23	$^3F_4 \to {}^3H_6$
8005,9	1,5487	1 900	+13	−37	$^3F_4 \to {}^3H_6$
8011,1	1,5477	340			$^3F_4 \to {}^3H_6$
8013,6	1,5472	950			$^3F_4 \to {}^3H_6$
8016,2	1,5467	3 750	+16	−18	$^3F_4 \to {}^3H_6$ $^1G_4 \to {}^3H_5$
8017,8	1,5464			−33	$^3F_5 \to {}^3H_6$
8021,4	1,5457				
8023,8	1,5452	11 760	+22	−45	$^3F_4 \to {}^3H_6$
8027,7	1,5445			−45	$^3F_4 \to {}^3H_6$
8029,5	1,5441	5 040	+20	−24	$^3F_4 \to {}^3H_6$
8032,1	1,5436	6 400	+10	−12	$^3F_4 \to {}^3H_6$
8033,8	1,5433	3 520		+69	$^3F_4 \to {}^3H_6$
8037,4	1,5426	86			
8043,4	1,5414	252	+18		$^3F_4 \to {}^3H_6$ $^1G_4 \to {}^3H_5$
8047,9	1,5406	168	+33	+23	$^3F_4 \to {}^3H_6$
8051,6	1,5399	560	+20	+17	
8061,7	1,5380	60		+75	$^3F_4 \to {}^3H_6$ $^1G_4 \to {}^3H_5$
8071,9	1,5360	224		+80	$^3F_4 \to {}^3H_6$ $^1G_4 \to {}^3H_5$
8093,1	1,5320	50		−60	$^3F_4 \to {}^3H_6$ $^1G_4 \to {}^3H_5$

1 Plus sign corresponds to quenching of the given intensity of infrared radiation, and a minus sign corresponds to an increase or build-up.

2 Plus corresponds to a predominant direction of the radiation E-vector parallel to the optical axis of the crystal

• The position of these lines was determined at 15 K since they flare up and are not visible still at 4.9 K. Unfilled entries correspond to a lack of data

Table 3. (Continued)

Wavelength λ in Å	Photon Energy $h\nu$ in eV	Intensity in arbitrary units	Quenching Coefficient[1] K, in %	Polarization[2] P in %	Transitions
8099,5	1,5308	50			$^3F_4 \to ^3H_6$
8115,0	1,5278	10			$^3F_4 \to ^3H_6$
8139,5	1,5232	77			$^3F_4 \to ^3H_6$
8152,7	1,5208	7			$^1G_4 \to ^3H_5$
8164,3	1,5186	34			
8181,7	1,5154	25			
8203,4	1,5114	5			$^1G_4 \to ^3H_5$
8225,9	1,5072	14			$^1G_4 \to ^3H_5$
8263,2	1,5004	4			

radiative transition from a given excited term sublevel to all lower-lying sublevels is comparable or even higher than the probability of radiationless transition from that sublevel to all remaining sublevels of the same term. In addition, the population of some of the higher sublevels may be maintained via the energy of thermal motion. From this it follows that in the emission spectrum there may be some lines present that are connected, generally speaking, with transitions from some upper term sublevel to some lower. In this way we derive that the possible number of lines in this region of the spectrum is three times more than observed. If, moreover, different luminescence centers were to yield lines that were also different in terms of their location, then for three kinds of centers we would obtain in sum more than two thousand lines and they would be so densely spaced in the emission spectrum that, even if they were 1 meV wide, they would form one continuous band. Since this was not obtained in our experiments, we may consider that the luminescence centers are differentiated basically not by position, but only by the relative intensity of lines (and, of course, by the kinetic parameters, which affect the spectrum only indirectly, through their contribution to one or another center).

However, small differences in the position of lines belonging to different centers did, after all, appear in the experiment. They can be noted if the line systems in the spectra of ZnS—Tm samples prepared under different conditions are compared. Although the accuracy of measuring the position of individual lines was no less than 0.2 Å, their positions coincided for different samples only to within 1 Å and sometimes only to within 1.5 Å. The differences in the line intensity of different samples was also very noticeable. In some of the samples individual lines might even be completely absent (or, more precisely, disappear below the overall noise level). Thus the fine structure lines in the blue end of the of the ZnS—Tm luminescence spectrum are noticeably fewer than in ZnS—Tm,S.

The fact that different centers make various contributions to different spectral lines is also apparent from differences in the value and even in the sign of the infrared sensitivity of close lines (a plus sign indicates quenching and a minus indicates a build-up of the line from the action of infrared radiation). And at the same time the radiators in these centers are strictly oriented in space. This proceeds from the high degree of polarization of many of these lines, reaching 50% and more. The field created by hexagonal interlayers in the crystal may be the factor which orients the radiators in these centers. Differences also in the sign of the polarization may be simply connected with the fact that each line corresponds to its own elemental radiator which may or may not be linked to the orientation of the other radiators.

AN APPROACH TO IDENTIFICATION OF THE SPECTRUM

To identify the spectrum means to find the sublevel system of terms, between which transitions occur that are responsible for that spectrum. If the spectrum is complete enough and the position of the lines are known fairly accurately, then almost all the information needed for this interpretation is already contained in the spectrum itself. In this case, for identification we may utilize the fact that the distances between lines (on the energy scale) arising during transitions between a certain sublevel of an upper term and all the lower sublevels accurately indicate the distance between the lower term sublevels, independent of which of the upper term sublevels is participating in these transitions. Each of these gives its own group of lines with the same resulting distance between lines. Thus for deriving the sublevel system it is enough to break up the totality of the lines into such groups having identical spacing inside the group. Unfortunately however, in the arguments we have presented it is possible for the upper and lower terms to swap places and not violate the reasoning. This means that it is not possible to establish from spectral data alone which of the found sublevel systems relate to the upper and which to the lower term. We have to use additional information to do this, e.g. the number of sublevels into which one or another term may be split (if we are given to understand that the number of these sublevels will be different for the upper and lower terms).

In practice, however, if the approach we have described is applied in its, so to speak, pure form, it will turn out to be ineffectual, since already for a spectrum containing several hundred lines the probability of random coincidence in the distances between lines is quite high. In our case the situation is even worse since, as has already been noted, there are almost twice as many possible combinations of upper- and lower-term sublevels as are observed, and the positions of these same lines changes slightly from sample to sample. Thus it is necessary to find some sort of independent means with which we could determine the distances between even just a few of the sublevels of upper or lower terms. Besides which, the incompleteness of the spectrum as it is measured in the defined energy range makes it doubtful that we could compare the number of detected sublevels with the theoretically-allowed splitting of terms. This forces us to search for some other method allowing us to clarify by experiment whether or not a given line group is united by common upper or lower sublevels.

For this purpose we can use the fact that in afterglow the intensity time dependence of all lines arising during transitions from the same sublevel must be strictly identical, else it indicates a time dependence of the population of that sublevel. The differences in photon energies of these lines is equal to the spacing between sublevels of the lower term. It is possible to find the position of upper-term sublevels, if several such groups are found which are responding to transitions to the very same lower-term sublevel from various upper-term sublevels. However, if several upper-term sublevels are found to be in thermal equilibrium with one another (which may be the case for high enough probabilities of forward and reverse transitions between them), then the lines arising from transitions from any of them will decay identically. In this case it is not possible to establish which of these correspond to transitions from the very same sublevel.

We can obtain more information if there is no thermal equilibrium between sublevels. As our calculations showed, for certain specific ratios of the probabilities of transitions between levels the population of some of them may even grow for some time after cessation of pumping. Correspondingly the lines arising during transitions from these sublevels will have a maximum in intensity. In some cases, two maxima are possible. Of course, if there are two lines having such an unusual afterglow, and moreover they are identical in this, then one may consider that they arise during a transition from the very same sublevel. However, it is not likely that there will be very many such lines in the spectrum. If we are to be satisfied with the less notable deviations from exponential decay, we will have to measure them with great precision, at least down to two or three figures in intensity, and to about 1 Å in widths of individual lines, a thing which is not trivial. Thus it is better to utilize the temperature dependence of individual lines' intensities (at low temperatures) for identification. In fact, if two lines have identical temperature dependence, then they in all probability arise during transitions from the very same upper-term sublevel to different lower-term sublevels. From the activation energy of quenching one can judge the presence and location of higher sublevels of the upper term through which the quenching takes place. If, on the other hand, some lines build up in intensity with increasing temperature, this indicates that the corresponding sublevel is "fed" by lower-lying sublevels.

TWO UPPER-TERM SUBLEVELS

We will look into this question in some detail. For starters we will examine only two sublevels of the excited state and assume that the probability of transitions between them is much greater than the probability of transitions between these sublevels and the other remaining sublevels of the upper term (Fig. 14). The kinetic equations of such a system have the form

$$\frac{dn_1}{dt} = \alpha_1 - n_1 w_1 - n_1 w_{12} + n_2 w_{21},$$

$$\frac{dn_2}{dt} = \alpha_2 - n_2 w_2 - n_2 w_{21} + n_1 w_{12}, \tag{1}$$

where n_1 and n_2 are the concentrations of centers with the corresponding excited states, and w_1 and w_2 are the summed probabilities of transitions from these upper-term sublevels to all lower-lying sublevels. Strictly speaking, the probability of radiationless transitions to these sublevels also enters into this, as well as radiative or radiationless transitions to all lower-lying sublevels of the upper term which are considered to be located so far down that thermal transition in the reverse direction is impossible. We will denote with w_{12} and w_{21} the probabilities of transitions between the investigated upper-term sublevels as

$$w_{2i} = w_{12}\exp\left(-\frac{\Delta E_{12}}{kT}\right),$$

(2)

where k is the Boltzmann constant, T is temperature, and ΔE_{12} is the energy difference between these sublevels. Finally, α_1 and α_2 are the pumping intensities of levels 1 and 2 (i.e., the number of excitations taking place per unit time per unit volume). By what means this excitation takes place — whether the result of direct absorption of pumping radiation photons corresponding to transitions to these levels (if stimulated emission is neglected), or during transitions from some higher state for which return by thermal means would be impossible, or even by some other route — is immaterial in this situation.

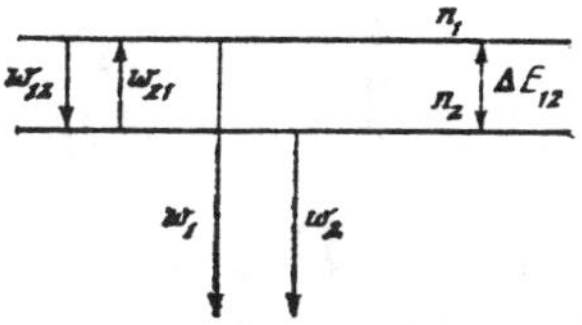

Fig. 14. Diagram of Transitions and Thermal Quenching in the Two Sublevel System

w_{ij} — Transition probability

n_i — Population of sublevels

ΔE_{12} — Energy difference between them

In the steady-state case, both differentials are equal to zero and Eq.(1) is transformed into a linear algebraic equation. Substituting in Eq.(2), it is not hard to solve for n_1 and n_2 and find thereby the temperature dependence of the concentration of centers excited up to sublevels 1 and 2. Multiplying them by the probabilities $w_{1изл}$ and $w_{2изл}$ of radiative transition from these sublevels to some specific lower-term sublevel, we find the temperature dependence of intensities I_1 and I_2 of the corresponding lines during constant pumping intensity:

$$I_1(T) = I_1(0) + \frac{B}{\dfrac{1}{A}\exp\left(+\dfrac{\Delta E_{12}}{kT}\right)+1}, \quad I_2(T) = \frac{I_2(0)}{A\exp\left(-\dfrac{\Delta E_{12}}{kT}\right)+1},$$

(3)

where

$$I_1(0) = \frac{\alpha_1 w_{1изл}}{w_1 + w_{12}}, \quad I_2(0) = \frac{w_{2изл}}{w_2}\left(\alpha_2 + \alpha_i \frac{w_{12}}{w_1 + w_{12}}\right),$$

$$A = \frac{w_1}{w_2}\frac{w_{12}}{w_1 + w_{12}}, \quad B = \frac{w_{1изл}}{w_1}\left(\alpha_2 + \alpha_i \frac{w_{12}}{w_1 + w_{12}}\right) \equiv I_2(0)\frac{w_2 w_{1изл}}{w_1 w_{2изл}}.$$

(4)

Constants $I_1(0)$ and $I_2(0)$ have the physical meaning of being the intensity of the corresponding lines at low enough temperatures that thermal transitions between sublevels can be neglected. As is apparent from Eq.(3), the value of I_1 grows upon increase in T, since the exponential argument in the denominator is positive, and I_2 decreases since the analogous argument is negative.

In the expression for $I_2(T)$ there are only three constants (counting the scale factor $I_2(0)$). We may algebraically find these from the experimental data, knowing the luminescence intensity measured in arbitrary units at three temperatures. First of all for this we must eliminate the scale factor,

93

normalizing the intensities at the point corresponding to the lowest temperature:

$$\frac{I_2(T_2)}{I_2(T_1)} = \frac{1 + A\exp\left(-\dfrac{\Delta E_{12}}{kT_1}\right)}{1 + A\exp\left(-\dfrac{\Delta E_{12}}{kT_2}\right)}, \qquad \frac{I_2(T_3)}{I_2(T_1)} = \frac{1 + A\exp\left(-\dfrac{\Delta E_{12}}{kT_1}\right)}{1 + A\exp\left(-\dfrac{\Delta E_{12}}{kT_3}\right)}. \tag{5}$$

Getting rid of the denominator in both equalities and collecting the A terms on the right-hand sides we obtain

$$I_2(T_1) - I_2(T_2) = A\left[I_2(T_2)\exp\left(-\frac{\Delta E_{12}}{kT_2}\right) - I_2(T_1)\exp\left(-\frac{\Delta E_{12}}{kT_2}\right)\right],$$

$$I_2(T_1) - I_2(T_3) = A\left[I_2(T_3)\exp\left(-\frac{\Delta E_{12}}{kT_3}\right) - I_2(T_1)\exp\left(-\frac{\Delta E_{12}}{kT_1}\right)\right]. \tag{6}$$

Dividing these equalities by one another we eliminate that very constant A. As a result we obtain an equation in terms of a single unknown ΔE_{12}:

$$\exp\left(\frac{\Delta E_{12}}{kT_1} - \frac{\Delta E_{12}}{kT_3}\right) = \frac{[I_2(T_3) - I_2(T_2)]\,I_2(T_1)}{[I_2(T_1) - I_2(T_2)]\,I_2(T_3)} +$$
$$+ \frac{[I_2(T_1) - I_2(T_3)]\,I_2(T_2)}{[I_2(T_1) - I_2(T_2)]\,I_2(T_3)}\exp\left(\frac{\Delta E_{12}}{kT_1} - \frac{\Delta E_{12}}{kT_2}\right), \tag{7}$$

which may be solved numerically. After this, with the help of any one of the equalities in (6) we may solve for A, and then from (4) $I_2(0)$, also in the same arbitrary units. After this, as a check, we can calculate $I_2(T)$ for the other temperature points and compare the results obtained with experimental measurements. For another, independent, confirmation of the correctness of ΔE_{12} we may use this variable itself. Adding it to the photon energy of an investigated line, we obtain the photon energy of the line generated during transitions from the upper sublevel of the excited state to the same sublevel of the ground state. If there is such a line in the luminescence spectrum of the given sample, then this serves as confirmation to the correctness of our method of finding ΔE_{12}.

To find ΔE_{12} from the temperature dependence of I_1 is more complicated, since here we have to find four instead of three constants. If the constant $I_1(0)$ were known, then deducting it from all the values of $I_1(T)$ we would obtain a quantity dependent on temperature in an analogous way to the Mott formula, with the only difference being that the exponent would be positive, not negative. In this case we could utilize the procedure for finding ΔE_{12} described above, and the ΔE_{12} we obtain would be negative. However, since $I_1(0)$ is not known, we have to find it (and the other constants too) by successive approximation. As a zeroth-order approximation for $I_1(0)$ we will take the value of I_1 measured at the lowest temperature. Then at the three points of correspondingly higher temperature we will find ΔE_{12}, A and B by the method described above, and with them calculate what I_1 should be at that temperature corresponding to $I_1(0)$. If this quantity is higher than the experimental value by more than the measurement error, we will choose a correspondingly lower value for the new $I_1(0)$, and so on. The accuracy with which we find ΔE_{12} in this case is, however, much lower, since due to the presence of three independent constants (not counting the scale factor), the approximate nature of the formula is more significant. Thus random scatter of the experimental points may strongly affect the results.

94

THREE UPPER-TERM SUBLEVELS

We turn now to a system with three sublevels in the upper term. For simplicity we will assume that the transition probability between the first and third sublevels, passing by the second, is negligibly small. Then the system of kinetic equations will have the form:

$$\frac{dn_1}{dt} = \alpha_i - n_1 w_i - n_1 w_{12} + n_2 w_{2i}',$$

$$\frac{dn_2}{dt} = \alpha_2 - n_2 w_2 - n_2 w_{23} - n_2 w_{2i} + n_1 w_{12} + n_3 w_{32}, \qquad (8)$$

$$\frac{dn_3}{dt} = \alpha_3 - n_3 w_3 - n_3 w_{32} + n_2 w_{23},$$

where

$$w_{2i} = w_{12} \exp\left(-\frac{\Delta E_{12}}{kT}\right), \qquad w_{32} = w_{23} \exp\left(-\frac{\Delta E_{23}}{kT}\right). \qquad (9)$$

The system specified here (Fig. 15) is the same as in the case of two sublevels: a single subscript indicates that the given quantity refers only to one sublevel, and two subscripts refers to a transition between two levels, the number of the initial sublevel being written first.

As before, setting the differentials equal to zero and substituting Eq.(9) into the right-hand side of Eq.(8) we again obtain a system of linear algebraic equations which we may solve for n_1, n_2 and n_3. As a result we obtain

$$n_i = \frac{C + D \exp\left(-\dfrac{\Delta E_{12}}{kT}\right) + BC \exp\left(-\dfrac{\Delta E_{23}}{kT}\right) + G \exp\left(-\dfrac{\Delta E_{12} + \Delta E_{23}}{kT}\right)}{1 + A \exp\left(-\dfrac{\Delta E_{12}}{kT}\right) + B \exp\left(-\dfrac{\Delta E_{23}}{kT}\right) + F \exp\left(-\dfrac{\Delta E_{12} + \Delta E_{23}}{kT}\right)}, \qquad (10)$$

$$n_2 = \frac{\alpha_2 + \alpha_3 \dfrac{w_{23} \exp\left(-\dfrac{\Delta E_{23}}{kT}\right)}{w_3 + w_{23} \exp\left(-\dfrac{\Delta E_{23}}{kT}\right)} + n_1 w_{12}}{w_2 + w_{12} \exp\left(-\dfrac{\Delta E_{12}}{kT}\right) + \dfrac{w_3 w_{23}}{w_3 + w_{23} \exp\left(-\dfrac{\Delta E_{23}}{kT}\right)}}, \qquad (11)$$

$$n_3 = \frac{\alpha_3 + n_2 w_{23}}{w_3 + w_{23} \exp\left(-\dfrac{\Delta E_{23}}{kT}\right)}, \qquad (12)$$

where

$$A = \frac{w_1 w_2}{(w_1 + w_{12})(w_2 + w_{23})}, \qquad B = \frac{w_{23} w_2}{w_3 (w_2 + w_{23})}, \qquad C = \frac{\alpha_1}{w_1 + w_{12}}, \qquad (13)$$

$$D = \frac{[(\alpha_1 + \alpha_2) w_{12}}{(w_1 + w_{12})(w_2 + w_{23})}, \qquad F = \frac{w_1 w_{12} w_{23}}{w_3 (w_1 + w_{12})(w_2 + w_{23})},$$

$$G = \frac{[(\alpha_1 + \alpha_2 + \alpha_3) w_{12} w_{23}}{w_3 (w_1 + w_{12})(w_2 + w_{23})}.$$

These formulas contain so many unknowns that, given the existing experimental error, to find any kind of reliable values for the spacing between levels ΔE_{12} and ΔE_{23} would not seem to be possible. Nevertheless, by differentiation we may assure ourselves that n_1 always grows with increasing temperature, and n_3 decreases. The intensities of the corresponding lines also vary this way. Analysis of the temperature dependence of n_2 is significantly more difficult, and it cannot be carried out to the fullest extent. However, we may confirm that the function $n_2(T)$ does have a maximum under some conditions. Thus, if $\Delta E_{23} < \Delta E_{12}$;

$$w_3 \ll w_{23}\exp\left(-\frac{\Delta E_{23}}{kT}\right); \quad w_3 \ll w_2\exp\left(-\frac{\Delta E_{23}}{kT}\right) \text{ и } w_2 \ll w_{12}\exp\left(-\frac{\Delta E_{12}}{kT}\right),$$

then for $\alpha_1 \ll \alpha_3$ and $\alpha_2 \ll \alpha_3$ the value of $n_2(T)$ grows with increasing temperature for small T when all the exponents are small compared to unity, and on the other hand, when these exponents become close to unity $n_2(T)$ begins to decrease. (it is understood that the inequality is, in effect, only a sufficient but not necessary condition for a maximum of $n_2(T)$.)

Two further cases are of interest. If

$$w_1 \gg w_{12}, \qquad w_2 \gg w_3, \tag{14}$$

then for $\alpha_1 = \alpha_3 = 0$ (that is, for some sufficiently small α_1 and α_3):

$$n_2 \approx \frac{\alpha_2}{w_2 + w_{12}\exp\left(-\dfrac{\Delta E_{12}}{kT}\right)}, \tag{15}$$

$$n_3 \approx \frac{\alpha_2 w_{23}}{\left[w_2 + w_{12}\exp\left(-\dfrac{\Delta E_{12}}{kT}\right)\right]\left[w_3 + w_{23}\exp\left(-\dfrac{\Delta E_{23}}{kT}\right)\right]}. \tag{16}$$

In this case the intensity of lines generated during transitions from the second sublevel falls off, according to the Mott equation, with an activation energy equal to the distance from this sublevel to the first (i.e., the highest of the three). The intensity of lines generated by transitions from the third sublevel falls off according to an equation which is the product of two Mott formulas, with an activation energy equal to the distance between the first and second, and the second and third sublevels. Lines generated by transitions from the first sublevel are quenched by an even more complex equation. If, besides the inequalities in Eq.(14) the following inequality is also satisfied

$$w_2 \gg w_{23}, \tag{17}$$

then for $\alpha_3 \neq 0$ it can turn out that simultaneously

$$\alpha_2 \gg \alpha_3 \gg \alpha_2 \frac{w_{23}}{w_2 + w_{12}\exp\left(-\dfrac{\Delta E_{12}}{kT}\right)}. \tag{18}$$

Then both lines will be quenched in accordance with the Mott equation with an activation energy corresponding to the distance to the nearest sublevel lying above the one from which the transition starts which yields that line.

Fig. 15.   Diagram of Transitions and Thermal Quenching in the Two Sublevel System

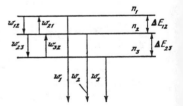

w_{ij} — Transition probability

n_i — Population of sublevels

ΔE_{12} — Energy difference between them

TEMPERATURE DEPENDENCE OF INTENSITY AND
THE FINAL PHASE OF IDENTIFICATION

We turn now to the experimental results. In the region around 2.6 eV corresponding to the $^1G_4 \rightarrow {}^3H_6$ or $^1D_2 \rightarrow {}^3H_6$ transitions the temperature dependence of intensity was studied for 15 lines, and the intensity was determined at all temperatures from the maximum of the line even when its position in the spectrum changed with changing temperature. It turned out that three lines had increases in intensity during heating, five passed through a maximum, and the remaining seven suffered a decrease. Examples of these three variations in temperature behavior are presented in Fig. 16. We note, incidentally, that we succeeded in observing temperature dependences of opposite "sign" — build-up and decay — for two very closely-spaced lines in the fine structure, differing by only 0.8 meV, as is evident from the data in Fig. 16. This testifies to the high resolution of our spectral temperature measurements.

Table 4 presents the corresponding experimental data as well as the value of ΔE calculated from them, where it has been possible to do so. Doubtful cases are denoted with a question mark. An example of such a case is also presented Fig. 16. In the other cases the probable error in determining ΔE is 0.1—0.2 meV. The minus sign for ΔE indicates that the given line is

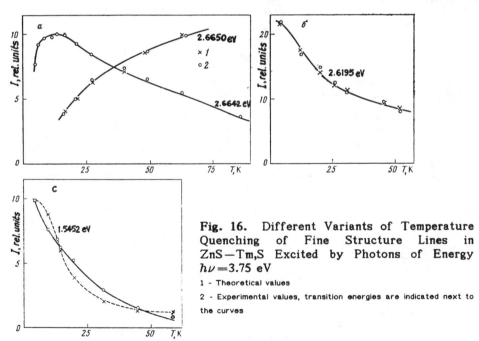

Fig. 16. Different Variants of Temperature Quenching of Fine Structure Lines in ZnS—Tm,S Excited by Photons of Energy $h\nu = 3.75$ eV

1 - Theoretical values

2 - Experimental values, transition energies are indicated next to the curves

generated by transition from the upper sublevel of a pair and a plus sign indicates that it is from the lower sublevel. Subtracting ΔE from the photon energy of the investigated lines in the first case and adding in the second, we can find the photon energy of lines generated by transitions from a different sublevel of a pair to the same sublevel of a lower term. The photon energies predicted by this method for emission lines are presented in the same Table ($h\nu_T$). By way of comparison, the photon energy closest to the predicted line is present in the next row, as it was found by experiment in the luminescence spectrum of ZnS—Tm at 4.9 K. As is apparent, the deviation in 7 of 10 cases is less than half an meV, i.e. it lies within the bounds of experimental error. In three of the cases we could not detect the corresponding lines (the difference between the predicted position of the line and the line closest to that in the spectrum exceeded by far the permissible measurement error for the photon energy). The absence of these lines in the measured spectra may be explained by the fact that in all three cases the position of the line was predicted from ΔE values that had not been very well determined.

As is apparent from Table 4, a distance between sublevels of close to 4.3 meV was repeated four times, and twice for something close to 2.7 meV. This can be interpreted as evidence that the corresponding lines are generated by transitions from the very same upper-term sublevel to a different sublevel of a lower term. From this we may find the distance between lower-term sublevels. A set of four sublevels was obtained with spacing of approximately 33.7, 2 and 14 meV and a pair of sublevels at a distance of 20.5 meV. In order to find the relative location of the set and pair of sublevels we studied the location of the remaining lines lying in the same region of the spectrum (for which the temperature dependence had not been studied). From these we found two pairs of lines ($h\nu=2.6172$ and 2.6126 eV and $h\nu=2.6369$ and 2.6326 eV) with a distance inside the pairs of

Table 4. Calculated (I_T) and Experimentally Measured ($I_э$) Temperature Dependence of Line Intensity in the Visible Spectrum With $h\nu=2.6$ eV.

λ, Å	4652,4		4653,8		4712,0		4717,7		4720,6		4723,2		4733,2	
hν, eV	2,6650		2,6642		2,6312		2,6281		2,6265		2,6250		2,6195	
I, K	I_T	$I_э$	I_T	$I_э$	I_T	$I_э$	I_T	$I_э$	I_T	$I_э$	I_T	$I_э$	I_T	$I_э$
4,7	—	—	7,7		—	—	8,5		7,7		122,5	122,5	21,9	22,0
13	—	—	10,0		1,0	1,0	9,7		9,6		112	95	17,8	17,0
16	3,9	4,0	9,9		—	—	—		—		—	—	—	—
20,5	5,0	5,0	9,3		3,7	3,5	9,8		8,2		89	80	14,0	15,0
26	6,2	6,1	8,5		5,9	5,0	9,6		6,1		75	75	12,3	12,5
31	7,1	7,0	7,8		7,8	7,5	9,0		4,9		66	64	11,2	11,0
46	8,7	8,7	6,5		12,4	13,0	8,2		2,8		50	55	9,4	9,5
62,5	9,7	10,0	5,7		15,7	16,0	6,3		—		42	42	8,4	8,0
85	—	—	3,6		—	—	3,2		—		—	—	—	—
ΔE, meV	−4,30		—		−4,33		—		—		4,27		2,74	
$h\nu_T$, eV	2,6607		—		2,6269		—		—		2,6292		2,6222	
$h\nu_э$, eV	2,6602		—		2,6265		—		—		2,6289		2,6224	

Question mark denotes a somewhat unreliable value of activation energy approximately 4.3 meV (characteristic for the splitting of upper levels, which is what generates the lines that allow us to find the four lower sublevels), and a distance between the pairs of around 20 meV, which corresponds to the second value for the activation energy found from the temperature dependence of the line intensity. Thanks to this connecting link we managed to find the location of six sublevels of the lower term and four of the upper term.

The distances thus found between the lower-term sublevels were then used to find the location of the remaining upper-term sublevels. Towards this end, we searched among the lines lying in this spectral region for a group having the same distance between, since this can be ascribed to transitions to the same lower-term sublevel. We found five such groups which, taken together with the four found earlier, corresponds exactly to a multiplicity of 9 for the $^1G_4$ term, while at the same time the $^1D_2$ term, and these lines may be ascribed to transitions starting from its sublevels, has a multiplicity of 5 and therefore can yield only five groups of lines. Table 5 presents the locations of emission lines calculated in accordance with the sublevel system as we found it, which are compared to the locations found experimentally. In composing this and all similar Tables following, in order not to introduce additional error the photon energy was calculated to five figures past the decimal and only then rounded off to conform with the limits of accuracy. Thus the difference between predicted photon energies for lines belonging to transitions from the same sublevel may have a remainder of one in the last digit. The predicted photon energies are given after allowance was made for the complete identifications, and so there might also be some differences with the values presented in Table 4.

In Table 5 the photon energy of lines whose temperature dependence we studied in the experiment are denoted with a solid underline, and those for which the locations were calculated from the activation energy with a

4735,8 2,6180		4748,2 2,6112		4756,8 2,6066		4766,0 2,6014		4769,3 2,5996		4781,9 2,5928		4782,6 2,5924		4941,6 2,5090	
I_T	$I_э$	I_T	$I_э$	I_T	$I_э$	I_T	$I_э$	I_T	$I_э$	I_T	$I_э$	I_T	$I_э$	I_T	$I_э$
12,5	12,5	7,0	7,0	1,0	1,0		4,0	49,8	50	22	22		8,6	6,0	6,0
11,5	10,0	6,4	6,5	1,3	1,5		13,5	40,5	40	21	15,5		10,1	5,7	5,0
—	—	—	—	—	—		—	—	—	—	—		—	—	—
9,9	9,6	5,1	5,0	3,1	3,5		15,5	31,9	32	12,5	10		9,0	5,3	5,0
9,0	9,0	4,3	4,5	4,7	4,5		15,5	28,0	28	7,0	7		8,0	5,0	5,0
8,4	8,5	3,8	4,0	5,9	5,5		15,5	25,5	22,5	4,4	5		6,0	4,8	5,0
7,2	7,5	2,9	3,5	8,2	8,5		15,0	21,5	21,5	1,7	2		4,0	4,3	5,0
6,0	6,5	2,4	2,5	9,2	9,0		12,0	19,2	14,5	1,0	1		2,2	4,0	4,0
—	—	—	—	—	—		—	—	—	—	—		0,8	—	—
3,33		4,27		−6,45 (?)		—		2,74		8,8 (?)		—		3,45 (?)	
2,6214		2,6155		2,6002		—		2,6023		2,6016		—		2,5124	
2,6218		2,6151		—		—		—		—		—		—	

dashed underline. As is apparent, the overwhelming majority of the predicted lines were also discovered by the experiment, and in columns 2 through 5, which were taken by us to be initial sublevels, no less than two of the lines are underlined each time, i.e. the corresponding energy difference obtained from the temperature dependence is no less than double. Possible error in the position of individual lines accumulates from measurement error (0.1 meV) and the scatter which we mentioned before in the position of lines from one sample to another (0.5—0.6 meV). The total may be considered to be 0.6—0.7 meV. However, the accuracy of measuring the distance between levels is greater than that, since it is found through optimal agreement with the predicted positions of several lines.

Table 5. Calculated ($h\nu_T$) and Experimentally Determined ($h\nu_э$) Values* of the Photon Energy (in eV) of Emission Lines Generated by Transitions From Sublevels of the $^1G_4$ Term to Sublevels of the $^3H_6$ Term

Sublevels of the Upper Term $^1G_4$

Sublevels of Lower Term $^3H_6$	0,0000	0,0175	0,0264	0,0448	0,0496	0,0527	0,0629	0,0672	0,0709
0,0549 [1]	$h\nu_T$ = 2,5428	2,5603	2,5692	2,5876	2,5924	2,5955	2,6057	2,6100	2,6137
	$h\nu_э$ = —	—	2,5697	2,5871	2,5924	2,5961	2,6061	2,6104	2,6135
0,0494	2,5483	2,5658	2,5746	2,5931	2,5979	2,6010	2,6111	2,6155	2,6192
	—	2,5655	2,5746	2,5928	2,5978	2,6014	2,6112	2,6151	2,6188
0,0478	2,5498	2,5673	2,5762	2,5946	2,5995	2,6025	2,6128	2,6171	2,6208
	2,5496	—	2,5767	2,5943	2,5996	—	2,6126	2,6172	2,6208
0,0415 [1]	2,5562	2,5737	2,5826	2,6010	2,6058	2,6089	2,6191	2,6234	2,6271
	—	—	2,5825	2,6014	2,6061	2,6083	2,6188	2,6235	2,6265
0,0362	2,5614	2,5789	2,5878	2,6062	2,6111	2,6141	2,6244	2,6287	2,6324
	2,5619	2,5793	2,5884	2,6066	2,6110	2,6144	2,6250	2,6289	2,6326
0,0336	2,5640	2,5815	2,5904	2,6088	2,6137	2,6167	2,6269	2,6312	2,6350
	2,5641	2,5816	2,5901	2,6083	2,6135	2,6172	2,6265	2,6312	2,6350
0,0280	2,5967	2,5872	2,5961	2,6145	2,6193	2,6224	2,6326	2,6369	2,6406
	2,5967	2,5871	2,5961	2,6144	2,6195	2,6224	2,6326	2,6369	—
0,0000	2,5977	2,6152	2,6241	2,6425	2,6473	2,6504	2,6606	2,6649	2,6686
	2,5974	2,6151	2,6235	—	2,6474	—	2,6602	2,6650	2,6686

* In Table 5, Tables 7 through 9 and 11 through 14 ,in each pair the value of the photon energy on top corresponds to theory and on bottom to experiment
1 These sublevels were found in a subsequent identification

UTILIZING THE DISTANCE BETWEEN SUBLEVELS FOR FURTHER IDENTIFICATIONS

The distances we have found between the $^1G_4$ and $^3H_6$ sublevels may be used for identification of the spectra corresponding to transitions from sublevels of the $^1G_4$ term to a lower-term sublevel, or from some other upper-term sublevel to a sublevel of the $^3H_6$ term. First of all this was done for the $^3F_4 \rightarrow {}^3H_6$ transition, since we also studied the temperature dependence of the intensity of a number of its lines. We also succeeded in finding the quenching activation energy ΔE, to be truthful, with satisfactory accuracy in only 5 of the 9 cases (Table 6). In the remaining cases the temperature dependence either had a maximum (whereupon ΔE, of course, was not calculated), or decayed, but not in accordance with the Mott formula. This is easy to see from the same Fig. 16 on the temperature dependence of the line intensity at 8023.8 Å (1.5452 eV), for which even the best value that could be selected, 4.4 meV, did not provide good agreement with experiment. The discrepancy between these curves is highly typical: if the theoretical curve is selected so that its start and end coincide with experiment, then in the middle it gives a significantly steeper dependence on temperature than is observed in the experiment. In similar situations the value of ΔE is indicated in the Table with two question marks. In all cases where we managed to find ΔE, we determined from it the position of lines paired to the investigated lines, and these lines were in fact detected in the spectrum.

The relative spacing of upper-term sublevels was determined in the following way. Each pair of lines was considered as generated by transitions to some certain lower-term sublevel, and from this assumption we calculated the position of lines corresponding to transitions from the same pair of upper sublevels to all remaining known sublevels of the lower term. Then we

Table 7. Calculated $(h\nu_T)$ and Experimentally Determined $(h\nu_э)$ Values of the Photon Energy (in eV) of Emission Lines Generated by Transitions From Sublevels of the Upper $^3F_4$ Term to Sublevels of the $^3H_6$ Term

Sublevels of Lower Term $^3H_6$	Sublevels of the Upper Term $^3F_4$								
	0,0000	0,0082	0,0155	0,0183	0,0209	0,0238	0,0248	0,0290	0,0320
0,0549¹	$h\nu_T = 1,5227$	1,5309	1,5382	1,5410	1,5436	1,5465	1,5475	1,5517	1,5547
	$h\nu_э = 1,5232$	1,5308	1,5380	1,5406	1,5436	1,5467	1,5472	1,5520	1,5547
0,0494	1,5282	1,5364	1,5437	1,5464	1,5490	1,5520	1,5529	1,5571	1,5601
	1,5278	1,5360	1,5433	1,5464	1,5487	1,5520	—	1,5568	1,5601
0,0478	1,5298	1,5380	1,5452	1,5480	1,5506	1,5536	1,5546	1,5587	1,5617
	—	1,5380	1,5452	1,5477	1,5509	1,5540	1,5547	1,5586	1,5620
0,0415¹	1,5360	1,5443	1,5516	1,5543	1,5569	1,5599	1,5609	1,5651	1,5681
	1,5360	1,5445	1,5520	1,5547	1,5575	1,5601	1,5612	1,5654	1,5677
0,0362	1,5414	1,5496	1,5568	1,5596	1,5622	1,5652	1,5662	1,5704	1,5734
	1,5414	1,5502	1,5568	1,5598	1,5620	1,5654	1,5661	1,5705	1,5728
0,0336	1,5440	1,5522	1,5594	2,5622	1,5648	1,5678	1,5688	1,5729	1,5759
	1,5441	1,5520	1,5591	1,5620	1,5646	1,5677	—	1,5728	1,5760
0,0280	1,5496	1,5578	1,5651	1,5678	1,5704	1,5734	1,5743	1,5786	1,5816
	1,5494	1,5577	1,5654	1,5677	1,5705	1,5728	1,5743	1,5783	1,5821
0,0000	1,5776	1,5858	1,5931	1,5958	1,5984	1,6014	1,6024	1,6066	1,6096
	1,5776	1,5857	1,5933	1,5957	1,5987	1,6015	1,6028	1,6072	1,6096

1 These sublevels were found in a subsequent identification

Table 6. Calculated (I_T) and Experimentally Measured ($I_э$) Temperature Dependence of the Intensity of Individual Lines in the Infrared Region of the Spectrum

λ, Å	7859,0		7867,1		7947,0		7948,8		7963,8		8005,9		8016,2		8023,8		8029,5	
hν, eV	1,5776		1,5760		1,5601		1,5598		1,5568		1,5487		1,5467		1,5452		1,5441	
T, K	I_T	$I_э$	I_T	$I_э$	I_T	$I_э$	I_T	$I_э$	I_T	$I_э$	I_T	$I_э$	I_T	$I_э$	I_T	$I_э$	I_T	$I_э$
5	9,5	9,5		10	9,8	9,8	9,0	9,0	10	10	10	10		8,7	10	10	10	10
10	9,5	9,5		10	9,8	9,8	9,0	9,0	9,6	9,3	9,2	9,0		10	8,9	7,7	8,4	7,6
15	9,2	9,0		9,9	9,4	9,6	8,8	9,0	8,8	8,6	7,8	8,0		9,1	6,0	6,3	4,4	5,6
20	8,1	8,2		9,6	8,4	8,1	8,3	8,6	8,0	8,0	6,5	7,0		7,4	3,9	5,3	2,3	4,0
32	4,8	5,0		8,0	5,0	5,0	6,9	6,8	6,8	7,0	4,7	5,0		4,4	2,0	2,9	0,9	0,9
45	2,8	2,9		5,6	3,1	3,2	5,7	5,7	6,1	6,8	3,8	3,8		—	1,3	1,4	—	—
59	2,0	1,6		2,3	—	—	—	—	—	—	—	—		—	1,1	0,7	—	—
ΔE, meV	8,2		—		7,9		6,4		3,0		3,4		—		4,4 (??)		5 (??)	
hν_T, eV	1,5858		—		1,5680		1,5662		1,5598		1,5521		—		—		—	
hν_э, eV	1,5857		—		1,5677		1,5661		1,5598		1,5520		—		—		—	

Question mark denotes a somewhat unreliable value of activation energy

Unfilled entries in the Table correspond to cases where there is a lack of experimental data, or where we failed to find conformity with theory

searched the emission spectrum for lines no more than 0.6 meV away from the prediction. After this the entire procedure was repeated with the assumption that the investigated line pairs are generated by transitions to some other sublevel of the lower term. For the true position of the lower sublevel it was undertaken to match up a great number of the lines which were close enough to the theoretical predictions. Since the approximate location of the lower sublevels was already established during analysis of the spectrum corresponding to the $^1G_4 \rightarrow ^3H_6$ transition, then refining the position of the $^3H_6$ term sublevels is possible only within the limits of $0.1-0.2$ meV, with the necessary checks to see that this does not downgrade the agreement with experiment obtained during the identification of the $^1G_4 \rightarrow ^3H_6$ transition. We succeeded in attaining unique identification with transitions in all four pairs of the investigated lines. The results of this identification are presented in Table 7. As is evident from the Table, out of 54 predicted lines in the spectrum we found 51, only failing to detect 3 of the lines.

It is interesting that the 0.0183 eV sublevel of the $^3F_4$ term appears as if in two hypostases — quenching for the 0.0155 ev sublevel and quenched through the 0.0248 eV sublevel. The possibility of such duality of effect by one and the same sublevel was indicated in the theoretical section of this article, so that we cannot consider this to be a contradiction in the identification. A second circumstance which would at first glance seem to contradict the identification is the fact that the 1.5601 eV line is generated by transitions from the quenching sublevel, not from the quenched sublevel as would follow from the temperature dependence. But this line is the most intensive in the spectrum and it is entirely possible that other, weaker lines with close photon energies and a different temperature dependence are being concealed beneath it. This very intensive line at 1.5601 eV is generated by transitions from the 0.0238 eV sublevel of the $^3F_4$ term to the 0.0415 eV sublevel of the $^3H_6$ term which was found after further identifications. In accordance with the temperature dependence of this line the 0.0238 eV sublevel may be quenched through the 0.0320 eV sublevel. The line corresponding to the transition from this sublevel to the same sublevel of the $^3H_6$ term is in the spectrum. This is why there is no contradiction with the temperature dependence of the line in this identification.

The splitting found for the $^3H_6$ term was utilized further for analyzing a group of lines belonging to the $^3F_3 \rightarrow ^3H_6$ transitions. Out of them only three groups of lines were found having a relative spacing corresponding to the splitting of the $^3H_6$ term, which allowed us to find the relative spacing of three sublevels of the $^3F_3$ term (Table 8). In this way the splitting we found for the $^3H_6$ term was corroborated by the locations of lines in three different radiative transitions (from the $^1G_4$, $^3F_3$ and $^3F_4$ terms).

Things turn out to be more complicated for splitting of the $^1G_4$ term. Four groups are found from among the lines belonging to transitions from this term to the $^3H_4$ term, which have a relative spacing corresponding to the five upper sublevels of the $^1G_4$ term found earlier, and yet we found not a single line corresponding to a transition from three lower sublevels of this term (Table 9). Moreover, the only line in this end of the spectrum ($h\nu = 1.9098$ eV) for which we studied the intensity temperature dependence turns out to be quenched with such a high activation energy (19.7 meV) that the corresponding sublevel must be 16 meV higher than the highest sublevel

103

Table 8. Calculated ($h\nu_T$) and Experimentally Determined ($h\nu_э$) Values of the Photon Energy (in eV) of Emission Lines Generated by Transitions From Sublevels of the $^3F_3$ Term to Sublevels of the Lower $^3H_6$ Term

Sublevel of Lower Term $^3H_6$	Sublevel of Upper Term $^3F_3$			Sublevel of Lower Term $^3H_6$	Sublevel of Upper Term $^3F_3$		
	0,0000	0,0022	0,0054		0,0000	0,0022	0,0054
0,0549[1]	$h\nu_T = 1,7241$	1,7263	1,7295	0,0362	1,7428	1,7451	1,7482
	$h\nu_э = 1,7241$	1,7257	1,7290		1,7434	1,7458	1,7481
0,0494	1,7296	1,7319	1,7350	0,0336	1,7454	1,7476	1,7507
	1,7290	—	1,7349		1,7458	1,7481	1,7509
0,0478	1,7312	1,7334	1,7366	0,0280	1,7510	1,7533	1,7564
	1,7310	1,7339	1,7366		1,7509	1,7537	1,7566
0,0415[1]	1,7371	1,7397	1,7429	0,0000	1,7790	1,7813	1,7844
	1,7366	1,7391	1,7434		1,7786	1,7812	1,7848

1 These sublevels were found in a subsequent identification

of the $^1G_4$ term. This is how we come up with a "pretended" 10 sublevels from 9 "places" between the sublevels of the $^1G_4$ term, and the existence of many of these were confirmed by various methods. On the other hand, the temperature quenching of the $h\nu = 1.9098$ eV line which agrees well with the Mott formula (Table 10) contradicts the temperature dependence of other lines ($h\nu = 2.6650$ and 2.6312 eV) generated by transitions from the sublevel of the $^1G_4$ term from which the transitions arise that must be ascribed to this line. From this it follows that the $h\nu = 1.9098$ eV line belongs to a transition

Table 9. Calculated ($h\nu_T$) and Experimentally Determined ($h\nu_э$) Values of the Photon Energy (in eV) of Emission Lines Generated by Transitions From Sublevels of the Upper $^1G_4$ Term to Sublevels of the Lower $^3H_4$ Term

Sublevels of Lower Term $^3H_4$	Sublevels of the Upper Term $^1G_4$								
	0,0000	0,0175	0,0264	0,0478	0,0496	0,0527	0,0629	0,0672	0,0709
0,0279	$h\nu_T = 1,8149$	1,8324	1,8413	1,8626	1,8646	1,8675	1,8778	1,8821	1,8858
	$h\nu_э = -$	—	—	—	1,8648	1,8678	1,8775	1,8825	1,8853
0,0099	1,8329	1,8504	1,8593	1,8806	1,8825	1,8855	1,8958	1,9001	1,9038
	—	—	—	1,8808	1,8825	1,8853	1,8959	1,8995	1,9043
0,0070	1,8358	1,8533	1,8622	1,8836	1,8854	1,8885	1,8987	1,9030	1,9067
	—	←	—	—	1,8853	1,8887	1,8982	1,9028	1,9068
0,0000	1,8428	1,8603	1,8692	1,8906	1,8924	1,8955	1,9057	1,9100	1,9137
	—	—	—	—	1,8926	1,8959	1,9053	1,9098	1,9137

Table 10. Calculated (I_T) and Experimentally Determined ($I_э$) Temperature Dependence of the Intensity of the $h\nu = 1.9098$ eV Line

T, K	I_T	$I_э$	T, K	I_T	$I_э$
4,9	1100	1100	79	190	190
16,4	1100	1090	108	97	71
26,2	1084	1060	138	63	64
37	930	920	183	43	46
55	465	480			

between some other terms. These terms could be $^3H_6$ too, since the distance between them is also close to 1.9 eV. The $h\nu = 1.9098$ eV line is the most intense in this end of the spectrum and can mask significantly weaker nearby lines having other temperature dependences (e.g., the $h\nu = 2.6650$ and 2.6312 eV lines). Therefore there is no basis for excluding this line from Table 9, although it is not possible to allow for its temperature dependence in this Table.

Table 11. Calculated ($h\nu_T$) and Experimentally Determined ($h\nu_э$) Values of the Photon Energy (in eV) of Emission Lines Generated by Transitions From Sublevels of the Upper $^3F_2$ Term to Sublevels of the Lower $^3H_6$ Term

Sublevel of Lower Term $^3H_6$	Sublevel of Upper Term $^3F_2$				
	0,0000	0,0046	0,0196	0,0253	0,0326
0,0549[1]	$h\nu_T = 1,8545$ $h\nu_э = -$	1,8591 —	1,8741 1,8744	1,8798 —	1,8871 1,8873
0,0494	1,8600 —	1,8646 1,8648	1,8796 —	1,8853 1,8853	1,8926 1,8926
0,0478	1,8616 —	1,8662 —	1,8812 1,8808	1,8869 1,8873	1,8942 1,8940
0,0415[1]	1,8679 1,8678	1,8725 1,8732	1,8875 1,8873	1,8932 —	1,9005 1,9009
0,0362	1,8732 1,8732	1,8778 1,8775	1,8928 1,8926	1,8985 1,8982	1,9058 1,9061
0,0336	1,8758 1,8761	1,8804 1,8808	1,8954 1,8959	1,9011 1,9009	1,9084 1,9084
0,0280	1,8814 1,8808	1,8860 1,8853	1,9010 1,9009	1,9067 1,9068	1,9140 1,9137
0,0000	1,9094 <u>1,9098</u>	1,9140 1,9137	1,9290 ┄┄ —	1,9347 —	1,9420 —

1 - These sublevels are found from a subsequent identification

The solid underlines denote lines for which the activation energy of quenching or build-up is known; the dashed underlines denote lines predicted in accordance with this activation energy

Table 12. Calculated $(h\nu_T)$ and Experimentally Determined $(h\nu_э)$ Values of the Photon Energy (in eV) of Emission Lines Generated by Transitions From Sublevels of the Upper $^1G_4$ Term to Sublevels of the Lower $^3H_5$ Term

Sublevel of Lower Term $^3H_5$	Sublevel of Upper Term $^1G_4$								
	0,0000	0,0175	0,0264	0,0478	0,0496	0,0527	0,0629	0,0672	0,0709
0,0246	$h\nu_T = 1,5114$	1,5289	1,5377	1,5592	1,5610	1,5641	1,5743	1,5786	1,5823
	$h\nu_э = 1,5114$	—	1,5380	1,5591	1,5612	1,5646	1,5743	1,5783	1,5821
0,0206	1,5154	1,5329	1,5418	1,5632	1,5650	1,5681	1,5784	1,5827	1,5864
	1,5154	—	1,5414	1,5628	1,5654	1,5677	1,5783	1,5831	1,5865
0,0157	1,5203	1,5378	1,5467	1,5681	1,5699	1,5730	1,5832	1,5875	1,5912
	1,5208	1,5380	1,5467	1,5677	1,5705	1,5728	1,5831	—	1,5914
0,0000	1,5360	1,5535	1,5624	1,5838	1,5856	1,5887	1,5989	1,6032	1,6069
	1,5360	1,5540	1,5620	1,5831	1,5857	—	1,5987	1,6028	1,6072

As will be explained further on, the intensive $h\nu = 1.9098$ eV line actually is part of a group which may be ascribed to transitions from one of the sublevels of the $^3F_2$ term to the five highest sublevels of the six found earlier for the $^3H_6$ (Table 11). In addition, lines of that same energy are part of a group belonging to the $^1D_2 \rightarrow {}^3F_3$ transition. Apparently however, in the $^3F_2 \rightarrow {}^3H_6$ this line is so much more intensive than its "twins" from the $^1G_4 \rightarrow {}^3H_4$ and $^1D_2 \rightarrow {}^3F_3$ transitions that its temperature dependence is not distorted by the presence of the other two lines. On the other hand, the relative spacing of the sublevels of the $^1G_4$ term were still corroborated by lines belonging to the $^1G_4 \rightarrow {}^3H_5$ transition. Four groups of lines were found, corresponding to transitions from these 9 sublevels, which made it possible to establish the relative spacing of 4 of the sublevels of the $^3H_5$ term (Table 12). This identification is confirmed by the fact that it also included a pair of lines with a spacing corresponding to the temperature dependence of one of them (as before, this line is highlighted with a solid underline in Table 11, and its pair is highlighted with a dashed line).

Having utilized the data we obtained on the splitting of the $^3H_5$, $^3F_3$ and $^3F_2$ terms, we may try out three independent methods for finding the sublevel system of the $^1D_2$ term, since transitions from it should yield lines which lie in the three regions of the spectrum which we've already investigated. Since our preliminary identification found in all a dozen lines belonging to these three terms, then we also searched for groups composed of a dozen lines with a strictly-defined spacing between each dozen and with the necessary condition that the distances between corresponding lines belonging to two pairs out of the dozen-line sets are equal to those found for one of the lines from the temperature dependence of intensity. We in fact managed to find five such dozen-line sets satisfying all the enumerated conditions. (True, two of them turned out to be incomplete — one of the dozen-line sets was missing a line, and another was missing three.) The consistency in the distance between these dozen-line sets was ascribed to splitting of the $^1D_2$ term (Tables 13, 14 and 15). As is apparent from these Tables, the location of each of the sublevels of this term are confirmed by no less than nine lines belonging to three different transitions.

An analogous device may be tried also for finding the missing sublevels of the of the ground term $^3H_6$. For this, we need to start from the already-found sublevels of the $^1G_4, {}^3F_2, {}^3F_3$ and $^3F_4$ terms from which transitions to $^3H_6$ arise that lie in the studied spectral regions. Since we found in all 23 sublevels belonging to these four excited terms, then transition from them to the same unknown sublevel of the $^3H_6$ term will yield a group of 23 lines with a relative spacing which we already know. Six such groups are

Table 13. Calculated $(h\nu_{\tau})$ and Experimentally Determined $(h\nu_{\vartheta})$ Values of the Photon Energy (in eV) of Emission Lines Generated by Transitions From Sublevels of the Upper $^1D_2$ Term to Sublevels of the Lower $^3H_5$ Term

Sublevel of Lower Term $^3H_5$	Sublevel of Upper Term $^1D_2$				
	0,0000	0,0026	0,0059	0,0108	0,0150
0,0246	$h\nu_{\tau} = 2,5912$	2,5938	2,5971	2,6020	2,6062
	$h\nu_{\vartheta} = 2,5914$	2,5943	2,5969	2,6014	2,6061
0,0206	2,5951	2,5977	2,6010	2,6059	2,6101
	—	2,5974	2,6014	2,6061	2,6104
0,0157	2,6000	2,6026	2,6059	2,6108	2,6150
	2,5996	—	2,6061	2,6110	2,6151
0,0000	2,6157	2,6183	2,6216	2,6265	2,6307
	2,6151	2,6180	2,6218	2,6265	2,6312

The solid underlines denote lines for which the activation energy of quenching or build-up is known; the dashed underlines denote lines predicted in accordance with this activation energy

Table 14. Calculated $(h\nu_{\tau})$ and Experimentally Determined $(h\nu_{\vartheta})$ Values of the Photon Energy (in eV) of Emission Lines Generated by Transitions From Sublevels of the Upper $^1D_2$ Term to Sublevels of the Lower $^3F_3$ Term

Sublevel of Lower Term $^3F_3$	Sublevel of Upper Term $^1D_2$				
	0,0000	0,0026	0,0059	0,0108	0,0150
0,0054	$h\nu_{\tau} = 1,8931$	1,8957	1,8990	1,9039	1,9081
	$h\nu_{\vartheta} = 1,8926$	1,8959	1,8995	1,9043	1,9084
0,0022	1,8963	1,8989	1,9022	1,9071	1,9113
	1,8959	1,8982	1,9028	1,9068	1,9112
0,0000	1,8985	1,9011	1,9044	1,9093	1,9135
	1,8982	1,9009	1,9043	1,9098	1,9137

Table 15. Calculated ($h\nu_T$) and Experimentally Determined ($h\nu_\mathfrak{g}$) Values of the Photon Energy (in eV) of Emission Lines Generated by Transitions From Sublevels of the Upper $^1D_2$ Term to Sublevels of the Lower $^3F_2$ Term

Sublevel of Lower Term $^3F_2$	Sublevel of Upper Term $^1D_2$				
	0,0000	0,0026	0,0059	0,0108	0,0150
0,0326	$h\nu_T = 1,7355$	1,7381	1,7414	1,7460	1,7505
	$h\nu_\mathfrak{g} = 1,7349$	—	1,7412	1,7458	1,7509
0,0253	1,7428	1,7454	1,7487	1,7536	1,7578
	1,7434	1,7458	1,7481	1,7537	1,7579
0,0196	1,7485	1,7511	1,7544	1,7593	1,7635
	1,7481	1,7509	1,7537	1,7590	1,7630
0,0046	1,7635	1,7661	1,7694	1,7743	1,7785
	1,7630	1,7662	1,7688	1,7747	1,7786
0,0000	1,7681	1,7707	1,7740	1,7789	1,7831
	1,7680	—	1,7747	1,7786	1,7828

already known. During spectral analysis we managed to find two more similar groups, although again they were incomplete. Five lines were missing in one, 3 in the other. If we assume the remaining 18—20 lines are satisfactory confirmation that these groups are not random, then we consider them as possible and place them in Tables 5, 7, 8 and 11, noting the corresponding sublevels with an asterisk. The remaining groups were significantly less complete, and so there was no basis for including them in the identification.

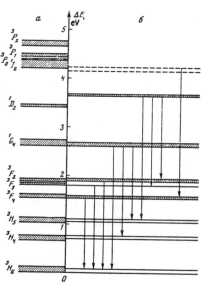

It was already noted above that during excitation of ZnS—Tm samples with photon energies exceeding and equal to 4.28 eV new lines were observed (cf. Fig. 8) which could be ascribed to transitions from a much higher-energy term than $^1D_2$. For clarification of this question we studied the spectra of ZnS—Tm during pumping by short wavelength radiation (at 77 K). To obtain photons with energy of 4.28 and 4.67 eV we used a pulsed dye laser with a doubling crystal, and a nitrogen laser for photon energies of 3.68 eV.

Fig. 17. Diagram of Tm^{3+} Terms of Observed Radiative Transitions

a - According to the data of [2]

b - Data from this article

The width of the band corresponds to the term splitting

Table 16 presents the energy-scale positions of lines observed during this pumping, averaged over all measurements during the different excitations. The accuracy of determining these lines was no less than 1 Å. From the data of the Table it is apparent that by throwing an electron at a higher term of Tm^{3+} we observe 7 new lines which are absent during pumping by 3.68 eV photons. We ascribed these lines to transitions from the $^1I_6$ level (we already remarked on the appearance of four of these, see Fig. 8, and their positions were measured earlier
with less accuracy). Judging from the photon energy, the seven lines we observed belong to the $^1I_6 \rightarrow {}^3F_4$ transition. As is apparent from Fig. 17, four of these in fact have relative spacing corresponding to the splitting of the $^3F_4$ term. Since this term is split into 9 sublevels, then transitions to them from some one sublevel of the $^1I_6$ term should give 9 lines. Four of the remaining five were also observed in the spectrum, although masked by lines belonging to other transitions (they are noted in the Table with an asterisk). We have as yet been unable to find one of the predicted lines. The remaining three lines appearing during short wavelength pumping are still unidentified. Apparently they belong to transitions from other $^1I_6$ term sublevels.

Table 16. Principal Lines in the Indigo Part of the Spectrum Excited by Photons of Various Energies at 77 K

Wavelength of Luminescence, Å	Energy of Excited Photons, eV	Pumping Photon Energy, in eV		
		3,68	4,28	4,77
4712,0	2,6312	+	+	+
4716,6	2,6287		+	+
4723,2	2,6250	+	+	+
4732,3	2,6200	+	+	+
4735,8	2,6180	+	+	+
4748,2	2,6112		+	+
4756,8	2,6065	+	+	+
4766,0	2,6014	+	+	+
4769,6	2,5995	+	+	+
4776,8	2,5956		+	+
4781,9	2,5928		+	+
4788,9	2,5890		+	
4793,7	2,5864		+	
4800,4	2,5828	+	+	+
4806,5	2,5795	+	+	+
4811,1	2,5771	+	+	+
4821,8	2,5713	+	+	+
4858,4	2,5520		+	+
4864,0	2,5490	+	+	+
4870,2	2,5458	+	+	+
4941,6	2,5090	+	+	+

The + indicates that a line is excited by the given photon

Table 17. Calculated ($h\nu_т$) and Experimentally Determined ($h\nu_э$) Values at 77 K of the Photon Energy of Lines Belonging to Transitions From the 0.0000 Sublevel of the Upper $^1I_6$ Term to a Sublevel of the Lower $^3F_4$ Term

Sublevel of Lower $^3F_4$ Term	$h\nu_т$	$h\nu_э$	Sublevel of Lower $^3F_4$ Term	$h\nu_т$	$h\nu_э$
0,0320	2,6112	2,6112	0,0183	2,5874	2,5871[1]
0,0290	2,6030	—	0,0155	2,5864	2,5864
0,0248	2,5957	2,5956	0,0082	2,5822	2,5825[1]
0,0238	2,5929	2,5928	0,0000	2,5792	2,5793[1]
0,0209	2,5903	2,5901[1]			

1 The position of these lines was found at 4.9 K. They are ascribed to different transitions.

Table 18. Splitting of Tm^{3+} Terms in ZnS—Tm and the Spacings Between Their Lowest-Lying Known Sublevels, According to [2] and the Results of This Article

Lower Terms	Upper Terms								Splitting
	$^3H_4$*	$^3H_5$*	$^3F_4$	$^3F_3$*	$^3F_2$	$^1G_4$	$^1D_2$	$^1I_6$	
$^3H_6$*	0,6962	1,0234	1,5567	1,8061	1,8675	2,5918	3,4333	4,2002	0,0988
	0,7548	1,0617	1,5776	1,7790	1,9094	2,5977	3,6557	4,2211	>0,0549
$^3H_4$*		0,3072	0,8605	1,1099	1,1713	1,8953	2,7371	3,5040	0,0712
		0,3069	0,8228	1,0242	1,1546	1,8429	2,9009	3,4663	>0,0279
$^3H_5$*			0,5333	0,7833	0,8441	1,5621	2,4099	3,1768	0,0816
			0,5159	0,7173	0,8477	1,5360	2,5940	3,1594	>0,0246
$^3F_4$				0,2494	0,3108	1,0348	1,8766	2,6435	0,0610
				0,2014	0,3318	1,0201	2,0781	2,6112	0,0320
$^3F_3$*					0,0614	0,7854	1,6272	2,3939	0,0197
					0,1304	0,8187	1,8767	2,4098	>0,0054
$^3F_2$						0,7240	1,5658	2,3325	0,0428
						0,6883	1,7463	2,2785	0,0326
$^1G_4$							0,8418	1,6085	0,1087
							1,0580	1,5902	0,0709
$^1D_2$								0,7667	0,0300
								0,5322	0,0150
$^1I_6$*									0,1692

* The sublevel system was not found completely. The underlined values are those found directly from the results of the identification

** In each pair of the presented values, the upper corresponds to data from [2], and the lower to data from this article.

110

We note further that the positions as found in this identification of the $^1D_2$ and $^1I_6$ terms agree with the results of [8], in which two transitions were observed that were not studied in this article. It was demonstrated there that the electroluminescence spectrum of ZnS—Tm,Cu crystals at room temperature had bands with maxima at 2.845 and 2.266 eV. In accordance with our identification, the first of these corresponds to the $^1D_2 \to {}^3H_4$ transition (and not to any other transition, if we take the location of the $^1D_2$ term as per [2]). That band with a maximum at 2.266 eV corresponds to the $^1I_6 \to {}^3F_2$ transition, according to our data.

The results of identification of the spectrum of Tm^{3+} in ZnS—Tm are cited in Table 18 and Figure 17. As is apparent from them, the spacings between the lowest-lying sublevels of all terms, except $^1D_2$, differ by no more than a few hundredths of an eV from those found in [2] for Tm^{3+} in Y_2O_3—Tm, and the splitting of all four terms (including $^1D_2$) for which all sublevels were found is less than the splitting in Y_2O_3—Tm. For those terms where not all the sublevels have been established, over the part that has been found the splitting occupies an interval 2 to 3 times smaller than that in Y_2O_3—Tm. Therefore we may consider that the overall splitting of these terms is also less in zinc sulfide than in Y_2O_3—Tm.

"DARK" PLACES IN THE IDENTIFICATION AND ADDITIONAL CHECKS ON IT

We will now discuss the several "dark" places in the identification. First, of the 236 lines which we located, it turns out that 70 of them are still not identified and it is unknown to what transitions between which sublevels they may be ascribed. Secondly, the total number of sublevels was found only in 4 of the 7 investigated terms. It would seem that we could ascribe the remaining unidentified lines to transitions involving these missing sublevels, however we did not manage to find an orderly system in this. Thirdly, of 340 predicted positions for lines, in 49 cases these lines did not turn up in the spectrum, and in many cases the very same line seemed to simultaneously belong to different transitions. It needs to be said that in these cases the predicted positions of lines differed little, so that the overlapping of two nearby lines was entirely possible, although in all the absence of a completely reciprocal unambiguous correspondence between predicted and experimentally detected lines did nothing to further our confidence in the uniqueness of the identification.

The splitting of the $^1D_2$ term is the least sound. It is substantiated by the temperature dependence of only one line, and in addition the average position of this term turned out to be roughly 0.2 eV higher than assumed in [2], which we made use of earlier [9]. Therefore an attempt was made to check that the identification of transitions from this term was not the result of a random "numbers game". With this in mind, an imitation spectrum made up of quasi-randomly-positioned lines was subjected to the analysis described above. We obtained this imitation spectrum in the following way. First, from the random number generator on a computer we obtained 100 numbers located with equal probability on the (0,1) interval. Then we arranged them in decreasing order, like the photon energies in the tables of real spectra. After this we counted out from the obtained sequence of numbers as many terms as there were lines in the spectrum we were imitating, and determined the range that they covered. All these numbers were then multiplied by the same coefficient, numerically equal to the range (in eV) occupied by the actual spectrum, in order to fill up that range with the numbers. All the remaining were also multiplied by this same coefficient; these numbers, as will become clear, act as a needed reserve during further operations. In order to account for the ultimate resolving power of the spectral measurements, pairs of "lines" in the imitation spectrum, located so close together that they could not be resolved in a real experiment, were replaced by a single "line" located exactly halfway between them. Since this reduced the number of "lines" in the imitation spectrum, we made up the necessary number of "lines" from the reserve, placing them immediately following the lines already included. This increases slightly the range covered by our imitation "spectrum", which makes it necessary to multiply all the numbers by a factor less than unity so that the range of the imitation spectrum will again correspond to the real one. Any pairs of "lines" that appear too close during this will again be excluded by the method described above and the whole procedure will be repeated. After two to three times of this we managed to obtain a set of not-so-random numbers that simultaneously satisfy three requirements: it contains as many lines as the spectrum it is imitating, it covers the same range and at the same time contains no lines with a distance between them less than the resolution of the spectral apparatus.

In the quasi-random "spectrum" thus obtained we managed to find no "lines" from which we could construct a table similar to Tables 13, 14 and 15. We succeeded in filling out, and with several omissions at that, three lines of a similar table, instead of five as in the case of a real spectrum. In other words, instead of finding four spacings between sublevels we found only two in the imitation spectrum. Of course, neither of these two spacings were equal to that which was confirmed from the temperature dependence of one of the lines in the actual identification. Besides which, no account was made for the known distance between lower terms in the "identification" of the model spectrum. This allowed us to move them slightly relative to one another, i.e. it gave us two free parameters which were not available in identification of the real spectrum. In spite of this, the "identification" of the model spectrum turned out to be substantially less complete than the identification of the real one.

However, some doubts about the correctness of our identification still remain. For example, we run into the energy difference 33—34 meV in the splitting of several of the terms, which is close to the energy of phonons in ZnS. There is no complete assurance that these actually belong to term splitting and not to phonon recurrence.

SOME RESULTS

We return to the beginning of this article and see what sorts of answers we have obtained to the questions posed there. Simplest of all is an answer to the question of the routes by which energy is received by luminescence centers. We explained that energy is passed by one of three routes, which we listed, and for the majority of terms the primary one is by resonant energy transfer from blue luminescence centers, which in turn get their energy by the recombination route. Only in the excitation of the $^1I_6$ term, which is located significantly higher than the level of the blue center, does the primary route turn out to be absorption of photons by the luminescence center itself. In the course of the investigation, however, a fourth route became apparent which we had not foreseen — two-stage excitation through an intermediate level of unknown origin.

The answer to the last question (about electronic transitions) also turned out to be more complicated than we presumed. We learned that the bands observed at room and nitrogen temperatures were in fact quite complex and simultaneously belonged to transitions between two and even three term pairs. Thus we had to give up the idea of an unambiguous comparison of transitions and the experimentally observed bands in Fig. 17.

Nonetheless, for a number of cases we did manage to establish, albeit after-the-fact, which terms were in fact being investigated through their transitions in the described experiments studying the kinetic properties of bands at room temperature. In these experiments, the spectral width of the monochromator slits that yielded the bands was, as a rule, no more than 3 nm. In the blue end this corresponded to approximately 16 meV, and in the very longest-wavelength regions to 10 meV. The investigated bands were 2 to 3 times wider. This means that, in fact, we investigated only the central part of a band, and not all the intensive lines fell within this. Unfortunately, at room temperature neither the precision of its position nor the relative intensity are known, since all lines are so broad that they overlap each other to a great degree and it is impossible to separate them. Therefore we had to use data obtained at 4.9 K , considering that external quenching will not develop at lower than room temperatures, and changes in the intensity of individual lines are only the result of competition between radiative transitions from various sublevels of the same term and do not change the overall number of radiative transitions. Assuming that, when the monochromator is set on the maximum of a band, only those lines lying no further than the spectral width of the slit away from the maximum (at 4.9 K) can contribute substantially to the measured signal, then we can make use of Tables $1-3$ and establish which transitions fundamentally contribute to one or another band.

In accordance with Table 1, band E_1 with a maximum at $h\nu_{max}=2.59$ eV is made up of the line at 2.5924 eV with intensity (at 4.9 K) of 7800 arbitrary units (henceforth the intensity will be indicated in parentheses after the photon energy), lines at 2.5901 (230) and 2.5871 (260) eV belonging to the $^1G_4 \rightarrow {}^3H_6$ transition, as well as the line at 2.5914 eV (450) belonging to the $^1D_2 \rightarrow {}^3H_5$ transition, and an unidentified line at 2.5888 eV (490). In this way about 90% of the E_1 band belongs to the $^1G_4 \rightarrow {}^3H_6$ transition.

The E_2 band with $h\nu_{max}=1.91$ eV, as per Table 2, is made up of three term-pair transitions which are comparable in terms of their contributions to the band: $^1G_4 \rightarrow {}^3H_4$, $^1D_2 \rightarrow {}^3F_3$ and $^3F_2 \rightarrow {}^3H_6$. It is made up of the line at 1.9003 eV (340) belonging to the $^1G_4 \rightarrow {}^3H_4$ transition, the line at 1.9112 eV (360) belonging to the $^1D_2 \rightarrow {}^3F_3$ transition, and three lines at 1.9098 eV (1100), 1.9137 eV (580) and 1.9068 eV (140) at once ascribable to all three transitions. However, the $^3F_2 \rightarrow {}^3H_6$ transition makes the basic contribution to the most intense of these (1.9098 eV). Therefore we must simultaneously ascribe all three of these transitions to the E_2 band.

The E_3 band ($\hbar\nu_{max} = 11.77$ eV), as per Table 2, is formed by the intensive lines at 1.7747 eV (1600), 1.7688 eV (220), 1.7680 eV (80), 1.7630 eV (810) and 1.7590 eV (90) generated by the $^1D_2 \rightarrow {}^3F_2$ transition, as well as unidentified lines at 1.7772 eV (90), 1.7650 eV (230) and 1.7608 eV (60), which made significantly less of a contribution to the overall intensity. Therefore, even if they do belong to some other transition, the E_3 could still and all be considered as belonging only to transitions from the $^1D_2$ term.

The fundamental contribution to the E_4 band ($\hbar\nu_{max} = 1.59$ eV), as per Table 3, is made by lines at 1.5914 eV (310) and 1.5831 eV (45) belonging to the $^1G_4 \rightarrow {}^3H_5$ transition, and to a lesser degree by the line at 1.5933 eV (42) generated by the $^3F_4 \rightarrow {}^3H_6$ transition and the line at 1.5821 eV (94) ascribed at once to both transitions. In addition, a slight contribution may be made from the intensive lines located at the limits of the investigated spectral region at 1.5783 eV (250), 1.5776 eV (2800) and 1.5760 (2300) belonging to the $^3F_4 \rightarrow {}^3H_6$ transition. On the basis of this data alone, the E_4 band should be considered as belonging to two transitions, however, as will be shown later, it apparently after all chiefly belongs to transitions from the $^1G_4$ term.

The E_5 band ($\hbar\nu_{max} = 1.54$ eV), as per that same Table 3, contains the largest grouping of intensive lines at 1.5452 eV (12,000), 1.5441 eV (5000), 1.5436 eV (6400), 1.5433 eV (3500) and 1.5406 eV (168) ascribed to the $^3F_4 \rightarrow {}^3H_6$ transition, as well as lines at 1.5415 eV (252), 1.5380 eV (60) and 1.5360 eV (224) attributed at once to two transitions. Their contribution, however, is less than 2% of the overall intensity of the band, so that it may be ascribed strictly to transitions from the $^3F_4$ term. However, as we noted earlier, the E_4 band differs markedly from the E_5 band in terms of its kinetic properties. It is therefore unlikely that transitions from sublevels of that same $^3F_4$ play any part in it. This is the basis for ascribing the E_4 band to transitions from the $^1G_4$ term. By the way, as was noted earlier, in terms of its kinetic properties the E_4 band has much in common with the E_1 band, which also belongs to the $^1G_4$ term.

The question of luminescence centers, as is ever the case, turns out to be most difficult. We have not been able to obtain any sort of detailed answers as of yet. On the basis of our data we can only say that thulium forms several structurally-different types of luminescence centers in zinc sulfide, and that the locations of these centers' sublevels differ by tenths of an eV, although the probability of transitions between them differs by a few hundred percent and possibly more.

CONCLUSIONS

Now it remains to indicate the course of future research. Its final goal is to clarify the structure of luminescence centers formed by trivalent ions of thulium in zinc sulfide. Similar centers are in all probability formed during doping by other trivalent elements.

After establishing the role of phonon recurrence in the luminescence spectrum and the corresponding refinement of the sublevel system of Tm^{3+}

terms, we should try to separate the spectra belonging to centers of different kinds. This demands careful investigation of the luminescence spectrum and polarization at helium temperatures measured with high spectral resolution. Since the structure of a Tm^{3+} center defines its kinetic parameters (the effective capture cross-section for charge carriers of the same and opposite sign and the location of the energy levels in the forbidden band) as well as the probability of receiving energy from a "blue" center, then it is possible to change the contribution to the overall luminescence spectrum from centers of a certain kind by changing the pumping conditions. By comparing one spectrum with another from the very same sample obtained during monochromatic pumping by photons that are of various energies at the instant of excitation and afterglow (in the range of seconds in order to tune out all intra-center processes), as well as under the action of infrared radiation with some certain photon energy, one can attempt to break up the overall spectrum into its constituent elements by utilizing the generalized Alentsev method [10]. We can also use the spectra of samples obtained under different preparation conditions for this. In particular, we can compare the spectra of an activator doped into a crystal during growth and the same activator doped electrolytically into an already-prepared crystal. Each constituent element, consisting of several lines, may be characterized by the fact that all of its lines change in intensity with changing pumping conditions in an identical fashion. It may be ascribed then to a center of some defined kind.

At the same time it also is necessary to investigate the spectral behavior of luminescence polarization during pumping by both polarized and unpolarized radiation. We should pay special attention to those lines which, according to the identification, belong to only a single transition. If it turns out that such lines are slightly displaced after a change in the pumping conditions, then this allows us to discriminate the contributions made to the line by centers of different kinds. Differences in polarization of the right and left halves of the line (of course, only if they can be detected) may also be of help in this. If we can detect a sensitivity in the polarization of a line to the polarization of the pumping radiation at one or another wavelength, then this will imply that the pumping radiation is directly absorbed by the luminescence centers, and these centers have a defined axis, which for each center is oriented along a specific crystallographic direction. For example, a Tm—O center in the corresponding lattice sites has an axis parallel to the <111> direction, while that of a Tm—cationic vacancy center is parallel to the <110> direction.

If even just a few lines for each term pair are found which can establish what contribution is made to the luminescence intensity by centers of different kinds and how they change with variation in pumping conditions, then the results obtained may be extended to the other lines by representing their intensity as a sum of the contribution of these centers. It is possible to check all this by comparing the spectra of samples obtained under different preparation conditions and therefore having different concentrations of these centers.

BIBLIOGRAPHY

1. Arkhangel'skiy, G. Ye., Bukke, Ye. Ye., Voznesenskaya, T. I., Grigor'yev, N. N. and Fok, M. V., "Visualization of the Structural Defects in ZnS-Type Crystals by the Anthrazyne Decoration Method" in the book "Metody vizualizatsii izobrazheniy" [Image Visualization Methods], Moscow, Nauka, 1981, pp 66-128 (also: TRUDY FIAN Vol 129)

2. Dieke, G. H., "Spectra and Energy Levels of Rare-Earth Ions in Crystals", New York, Wiley, 1968, 60 p

3. Ibuki, S. and Langer, D., "Energy Terms of ZnS:Tm and Ho", J. CHEM.PHYS Vol 40 pp 796-808, 1964

4. Yastrabik, L., Mares, I., Pachesova, S., Fok, M. V. and Grigor'yev, N. N., "Study of Non-Equivalent Radiative Tm^{3+} Centers in ZnS", J. LUMINESCENCE Vol 24/25 p 293-300, 1981

5. Gorbacheva, N. A., Grigor'yev, N. N., Pachesova, S., Fok, M. V. and Yastrabik, L., "Luminescence of ZnS—Tm Crystals" in the book: "Materialy XXVIII soveshch. po lyuminestsentsii kristallofosforov" [Proceedings of the Twenty-Eighth Conference on Crystal Phosphor Luminescence (Ezerniyeki, 13-16 May 1980): Report Theses], Riga, Institut fiziki AN LatvSSR, 1980, 224 pp

6. Charriere, Y. and Porcher, P., "Energy Levels and Crystal Field Parameters of ZnS:Tm^{3+}", J.ELECTROCHEMICAL SOC. Vol 125 pp 175-180, 1981

7. Grigor'yev, N. N., Ovchinnikov, A. V. and Fok, M. V., "Nontrivial Kinetics of Luminescence Polarization in ZnS Crystals", KRAT. SOOBSHCH. PO FIZIKE FIAN No 8 pp 25-30, 1981

8. L'vova, Ye. Yu. and Fok, M. V., "Collisional Excitation of Tm^{3+} During Electroluminescence in Crystals" in the book: "Tez. dokl. k rasshirennomu zasedaniyu sektsii elektrolyuminestsentsii Nauch. soveta po lyuminestsentsii AN SSSR" [Theses of Reports to the General Session of the Electroluminescence Section of the Scientific Council on Luminescence of the USSR Academy of Sciences (Tartu, 25-27 June 1985)], Tartu, Izdatel'stvo Tart. Univ., 1985, 21 pp

9. Yastrabik, L., Pachesova, S., Fok, M. V. et al, "Study of the Photoexcitation Mechanisms of Tm^{3+} Luminescent Centers in ZnS:Tm", CZECH. J. PHYS. B Vol 33 pp 1262-1271, 1983

10. Fok, M. V., "Breaking Down Complicated Spectral Bands Into Individual Bands With the Generalized Alentsev Method" in the book: "Lyuminestsentsiya i nelineynaya optika" [Luminescence and Nonlinear Optics], Moscow, Nauka, 1972, pp 86-135 (also TRUDY FIAN Vol 59)

Kinetics of Luminescence Polarization in Europium- and Thulium-Activated Single Crystals of Zinc Sulfide

N.N. Grigor'yev, A.V. Ovchinnikov,
M.V. Fok

Abstract: During research on afterglow in ZnS—Eu crystals it was discovered that the degree of luminescence polarization grows from 10 to 30% at room temperature over a period of about 10 μs after excitation by optical pulses 10 ns long with wavelength $\lambda=337$ nm, and this was valid for all three elemental bands, ascribed to Eu, which are the result of spectrum decomposition by the Alentsev method. Investigation of the temperature dependence of the rate of increase in the degree of polarization demonstrated that a potential barrier of about 0.37 eV is overcome during ordering of the orientations of the radiators. The degree of luminescence polarization in Zns—Tm afterglow, on the other hand, drops off from 20% to 0 in a few milliseconds. Both phenomena are explained by the fact that the hexagonal interlayer field in ZnS has a much stronger effect on RE^{3+} ions than RE^{2+}, forcing the RE^{3+} ions to move preferentially into one of four Jahn-Teller potential wells , which causes the preferred orientation of the radiators to be along the C-axis of the crystal. Differences in the sign of the effect are explained by the fact thet Eu is found in the Eu^{3+} state only until the luminescence center is ionized, whereas the Tm^{3+} ions are present in unexcited ZnS and convert to Tm^{2+} as a result of ionization of the luminescence centers.

INTRODUCTION

As is known from [1], spontaneous polarization of steady-state photoluminescence in ZnS:Eu single crystals reaches the tens of percentage points when the predominant direction of the electric field vector is along

117

the optical axis of the crystal (the C-axis). A high degree of polarization attests to a high degree of orientation among the radiators in the corresponding luminescence centers. It is still unclear, however, what the connection is with the fact that spontaneous luminescence polarization does not reach 100% even during excitation and observation perpendicular to the optical axis of the crystal, although the birefringence luminescence polarization at the azimuth of observation can have no effect on it, and the E-vector of the pumping radiation introduces no excitation anisotropy. Generally speaking, three reasons for incomplete polarization of luminescence light are possible. Either the situation is such that its spectrum is composed of a series of overlapping various-intensity bands with each neighboring bands in a pair having opposite-sign polarization, or the problem is a partial disordering of the radiators in the ensemble of luminescence centers, or finally it might be precession of each dipole radiator around a fixed axis (in this case, the C-axis).

For an explantion of which these causes plays the principal role, we started from the following considerations. If the first cause is the dominant one, then the degree of luminescence polarization will show a complicated spectral behavior with a number of extrema corresponding to the region of greatest contribution by one or another band to the overall luminescence spectrum. In this case, their relative contribution will change due to differences in risetime of the bands during afterglow. Therefore the degree of polarization of some certain spectral region of afterglow may grow gradually, and in others may decline or even change sign. For the second case the criterion could be the dependence of the degree of steady-state luminescence polarization on the azimuth of pumping radiation polarization (that is, if it is polarized, i.e. here we are speaking not of spontaneous, but rather of induced polarization), as well gradual decrease of the degree of polarization during afterglow caused by energy exchange between luminescence centers of different orientations. In contrast, the third case would be characterized by the fact that the luminescence polarization would be independent of this same azimuth of pumping radiation polarization and afterglow time.

SAMPLES AND METHODS

The investigated ZnS:Eu crystals contained on the order of 10^{-4} Eu (by weight). They were grown from the vapor phase in a closed ampule by the modified Griyo method [2]. Its luminescence spectrum consists of a series of highly overlapped bands. Therefore the first thing to do is to separate them into individual bands which are related to luminescence centers of the same kind. We used for this purpose the generalized Alentsev method [3]. This method was already employed for spectral analysis of ZnS:Eu in [4], but we were unable to take advantage of the results of that study since we required much greater accuracy (no less than 2%) for measurements and experimental data reduction to determine the degree of polarization.

In addition to a large number of measurements, spectral and polarization methods also require the introduction of several corrections for obtaining the final results. Moreover, the experiment was conducted using a standard algorithm. Therefore we developed a measuring/computation set-up

[5] based on an Elektronika-60 microcomputer with the standard peripherals and floppy-disk external storage to control the experiment and speed up reduction of the obtained data (Fig. 1). The CAMAC standard was utilized for connecting the physical subject with the computer. We primarily used the standard CAMAC modules, but also employed some nonstandard ones, specially designed for the specific nature of this job. The RT-11 operating system was installed on this computer. Programming for the experiment and preliminary data reduction was carried out using the QUASIC language, which is specially intended for CAMAC-standard micro- and mini-computers and measuring equipment. Version 1 of this language was expanded and loaded as a file into the OS. This allowed us to do without the inconvenience of punched tape and to store all the accumulated data on magnetic media.

A block diagram of the apparatus for measuring luminescence spectra and polarization during steady-state pumping is indicated in Fig. 2. In all our experiments the crystal was pumped perpendicular to the (110) plane. The optical axis of the crystal was oriented perpendicular to the slit of a UM-2 monochromator. We used as pumping radiation the $\lambda = 365$ nm Hg line separated out from the radiation spectrum of a PRK-4 mercury lamp attached to a stabilized power supply. A magnified image of the sample was projected on the monochromator slit with a Yupiter-9 objective. Since the investigated zinc sulfide crystal had a cubic structure with a significant amount of hexagonal phase impurity (up to 15%), this allowed us to select certain parts of the crystal for measurement which had about the same hexagonal phase content. The latter was determined from the distribution of color observed by passing white light through crossed polaroids. In order to decrease the polarization of the light induced by the apparatus we inserted a thin glass plate in the beam path before the monochromator input slit, which was rotated with respect to the monochromator optical axis at an angle such that the polarization of light reflected from this plate would in all probability compensate for polarization introduced by the remaining elements of the apparatus. Although complete correction of apparatus-induced polarization can only be managed at two points of the spectrum by this method, it is significantly reduced (by a few hundred percent) over the rest of the spectrum, which makes the corresponding corrections substantially easier. The wavelengths were discretely varied by rotating the monochromator drum with a DShI-200 stepping motor which was controlled by a specialized CAMAC module.

The FEU-79 optical detector operated in the photon counting mode. Cooling the FEU to $-15°$ C with a thermoelectric refrigerator lowered the level of thermal noise from 50 to 5 s^{-1}. Single electron pulses of 10 ns duration and 10 mV amplitude were passed on to a single-channel analyzer. In our analyzer, as opposed to the one in [7], the upper and lower threshholds were set by software which allowed us to select the optimal operating regime of the FEU while on-line with various samples. (The lower threshhold got rid of noise pulses while the upper was set to eliminate stray pickup pulses.) If a single electron pulse entering the FEU has an amplitude greater than the lower threshhold and less than the upper, then the analyzer generates a pulse with normalized amplitude. Furthermore it shapes a 100 ns TTL-standard logical pulse which is passed on to the binary counter connected to the CAMAC bus. The counter is controlled by a programmed timer which marks the time of the pulse. The spectrum was measured point-by-point, and

119

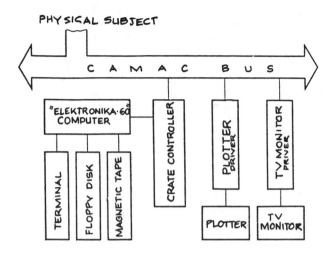

Fig. 1. Structural Diagram of Measuring/Computation Set-up for Optical Experiments

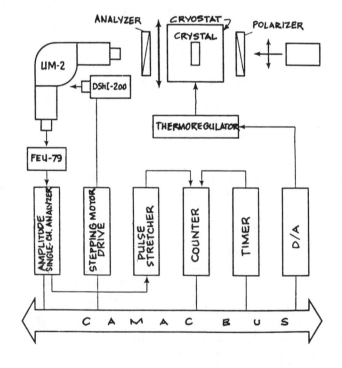

Fig. 2. Block Diagram of Apparatus for Measurement of Spectrum and Luminescence Polarization During Steady-State Excitation

this was output to a color graphics display[8] based on an ordinary color television and a specially-developed CAMAC module for on-line monitoring of both the counter contents and the statistical measurement error (inversely proportional to the square root of the number of recorded pulses). We amassed more than 10,000 pulses in the counter in order to obtain statistically reliable data with an accuracy of 1% or better. The counting period did not exceed 15 min for the lowest optical flux. The same method for 50 points in the spectrum allowed us to measure each with the same accuracy, which is important in the process of decomposing the spectrum into elemental bands.

After the last measurement is made, the results are output to the graphics display in the form of spectral curves, a visual check is made and, if there are no gross errors, the obtained data is reduced by the computer: corrections are made for apparatus-induced polarization and the spectral sensitivity function of the apparatus. After that the data is recorded and stored as files on magnetic tape. The data files containing all the correction factors are also located on that same disk. The polarization correction is determined for each wavelength by the ratio of two signals coming from a source of unpolarized light ; for one, the analyzer is set to pass only the E-vector perpendicular to the monochromator slit , and only the parallel E-vector is passed for the other. In this measurement the light source is an ordinary incandescent lamp, enclosed in frosted glass to cut down even that small amount of polarization arising from reflection of light on the walls of a light bulb. The correction for the spectral sensitivity function of the apparatus was determined with the aid of the known spectrum from an SI-40-100 lamp operating in the standard mode. Light from this lamp was incident on a screen made of freshly-deposited magnesium oxide placed in front of the monochromator where the crystal would usually be. Since the spectrum is given in units proportional to the number of photons per unit frequency interval, then the luminescence spectrum obtained after introducing correction for the spectral sensitivity of the apparatus is in these same units.

Final results of the experiment are documented, i.e. output in the form of curves on an x-y plotter and in the form of tables on a printer.

INVESTIGATION OF STEADY-STATE LUMINESCENCE

Our crystals were practically no different from the ZnS—Eu crystals investigated in [4] as regards impurity concentration and conditions of preparation. As one should expect, the results of investigation of their steady-state luminescence also turned out to be similar to those obtained earlier. It . was important to us, however, that even during careful measurement of luminescence polarization we detected no extrema in its spectral dependence (Fig. 3). This makes our first hypothesis very doubtful, but to refute it fully we need to make sure that there is not a single one of the luminescence bands, no matter how weak, which make up the spectrum all the way to $E = 2.4$ eV that is polarized with the opposite sign.

To find these individual bands and their contribution to the overall spectrum, as we said, we applied the generalized Alentsev method [3]. With this in mind, we measured four luminescence spectra during pumping with unpolarized light. Two of these were obtained for observations of luminescence intensity corresponding to directions for the light wave E-vector both perpendicular and parallel to the optical axis of the crystal at the same pumping intensity, and two more for the same E-vector directions but 10 times greater intensity. The spectrum thus normalized is presented in Fig. 4. As is evident from the Figure, the luminescence spectrum of ZnS:Eu depends on the pumping intensity. This testifies to the fact that the elemental bands making it up belong to different luminescence centers differing strongly in their kinetic parameters and, moreover, that they may indeed be used for separating out individual bands of the luminescence spectrum. The breaking up of the spectrum into its constituent parts was carried out interactively on a computer: the machine executed the mathematical computations and drew the curves on the plotter, and the operator found the segments over which the ratio of the four initial spectra are constant. In this way, we managed to separate out four elemental bands in all. The reliability of this method of generating bands was confirmed by the fact that the very same bands were obtained by solving this incorrectly-posed problem (in a mathematical sense) by various methods. Fig. 5 shows one of the initial spectra and its constituent elemental bands with maxima at $E_1=1.78$ eV, $E_2=1.96$ eV, $E_3=2.22$ eV, and $E_4=2.38$ eV, and halfwidths of 0.34, 0.3, 0.22 and 0.24 eV, respectively. The E_4 band, however, does not appear in all the spectra, and its origins are unclear.

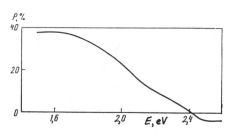

Fig. 3. Typical Spectral Dependence of Luminescence Polarization for ZnS:Eu Single Crystals With Steady-State Pumping by Unpolarized Light of $h\nu=3.39$ eV at 300 K

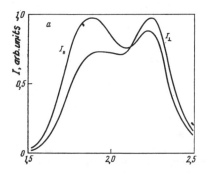
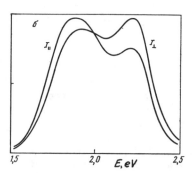

Fig. 4. Normalized Luminescence Spectra of ZnS:Eu (Sample No 1) With Steady-State Pumping by Unpolarized Light of $h\nu=3.39$ eV at 300 K, Used for Separating Out the Elemental Bands According to the Alentsev Method

$I_{\parallel}$ corresponds to the E-vector of the light wave parallel to the optical axis of the crystal, and $I_{\perp}$ corresponds to a perpendicular E-vector. In a the pumping intensity was 10 times greater than in b.

The last studies demonstrated that the contours of the elemental bands did not depend on the direction of the E-vector in the pumping radiation, which is the same for various ZnS:Eu samples and remained constant at various stages of afterglow during pulsed excitation. Therefore in the future the task of breaking down the spectrum amounts to determining the contribution of bands which are already known. The percent contribution is determined by the least squares method. The accuracy of the method is demonstrated by curve 6 of Fig. 5, which gives the difference ΔI between the measured spectrum and the spectrum synthesized from calculation of the

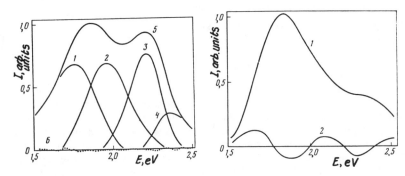

Fig. 5. Breaking Down the Luminescence Spectrum of ZnS:Eu Crystals (Sample No 2) Into Elemental Bands

Curves 1—4: constituent elemental bands with maxima at $E_1 = 1.78$ eV, $E_2 = 1.96$ eV, $E_3 = 2.22$ eV, and $E_4 = 2.38$ eV; 5 - One of the initial spectra for ZnS:Eu; 6 - Difference between spectrum 5 and one synthesized from the $E_1 - E_4$ bands, i.e. the discrepancy ΔI

Fig. 6. Illustrating That the Spectrum of a ZnS:Cu Crystal Cannot Be Broken Down Into the Elemental Bands for ZnS:Eu ($E_1 - E_4$)

1 - Initial ZnS:Cu spectrum; 2 - Difference between the initial spectrum and that synthesized from bands $E_1 - E_4$, for which the percent contribution was calculated by the least squares method.

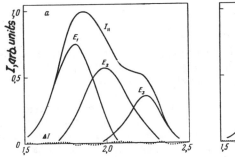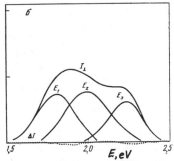

Fig. 7. Breaking Down the ZnS:Eu Spectrum (Sample No 3) Into Elemental Bands $E_1 - E_4$ by the Least Squares Method

$I_{\parallel}$ (a) and $I_{\perp}$ (b) are the luminescence spectra measured for two different positions of the analyzer

percent contribution of bands. As is evident, this difference is no more than 2 to 3% of the value of the ordinate of the initial spectrum and what is more, it changes sign often. Therefore we can attribute practically all of this to random scatter of points.

At the same time, if you attempt to break down the spectrum of the ZnS:Cu crystal, which is outwardly quite similar to the ZnS:Eu spectrum, into these same bands, the value of ΔI turns out to be about an order of magnitude greater. In addition it changes sign in a regular way, corresponding to the number of bands into which the spectrum is broken down in this test (cf. Fig. 6). In this way, a small ΔI (spectra discrepancy) is a reliable criterion for correctly breaking down of the spectrum into elemental bands. In the future we will use this criterion constantly.

The degree of luminescence polarization P is determined from the formula

$$P = \frac{I_\parallel - I_\perp}{I_\parallel + I_\perp} \ , \qquad (1)$$

where $I_\parallel$ and $I_\perp$ are, respectively, the intensities of luminescence light passing through an analyzer which transmits waves with an E-vector parallel and perpendicular to the optical axis of the crystal. In this manner $P > 0$ if the E-vector of the luminescence light is directed predominantly along the C axis. During the preliminary experiments, it was established what selected E-vector direction actually corresponds to an extremal value for intensity, since this is required by definition of the concept "degree of polarization" (incidentally, this was clear also from symmetry considerations, without experiments).

The investigation we conducted showed that all elemental bands except for the fourth had polarization identical in sign, although different in absolute value. The bands also differed greatly in their relative intensity, with the shortest-wavelength bands usually being the weakest. For example, the sample whose spectrum is shown in Fig. 7 has in all three bands. As is evident in the Figure, the longest-wavelength band (E_1) has the greatest degree of spontaneous polarization (34%), the middle band (E_2) has degree of polarization 19%, and the E_3 band — only 7%, which, incidentally, still lies far outside the bounds of experimental error.

The fact that the sign of the polarization is the same for all bands completely excludes the first hypothesis about the reason for depolarization of the luminescence from ZnS:Eu. As concerns the second and third hypotheses, study of steady-state luminescence gives an ambiguous answer. In fact, on the one hand, it turns out that, despite the constant sign of polarization of a band, its absolute value changes by a factor of $0.5-2$ from sample to sample. From the point of view of the second hypothesis, this is easy to explain if you assume that all bands belong to associative centers, which seem to have their orientation axes partially aligned during crystal growth. Nominally identical ZnS:Eu crystals can differ slightly in their preparation conditions, e.g. synthesis temperature, oxygen content or rate of cooling. These conditions can influence the degree of axis orientation in associative centers relative to the growth surface. Starting from the third hypothesis, it is hard to see how preparation conditions could affect the precession angle of radiators in luminescence centers. On the other hand, from the point of view of the second hypothesis, it is hard to explain the results of experiments where the luminescence was pumped with polarized light. In these experiments the luminescence was pumped with light normally incident to the surface of a plate cut from ZnS:Eu crystal so that the C axis was in the plane of the plate. In this situation, birefringence did not influence the polarization of the pumping radiation if the azimuth of the polarization plane was parallel or perpendicular to the optical axis of the crystal. Luminescence was observed by transillumination, and it turned out that, to within 2% accuracy the luminescence polarization was insensitive to the direction of the E-vector of the pumping radiation. But during polarized-light pumping, associative centers of a given orientation are preferred over the others, which inevitably must influence the degree of luminescence polarization. From the point of view of the third hypothesis, the luminescence polarization does not depend on the pumping polarization since the precession angle of the radiators in the luminescence centers also does not depend on it.

KINETICS OF POLARIZATION IN ZnS:Eu

The contradiction noted above in interpretation of the results of various experiments leads to the idea that we have overlooked some sort of essential fact in our arguments. This makes the kinetics experiments especially important as a source of new information about these very processes.

It is best to conduct measurement of the kinetics of luminescence polarization in the region of spectrum where overlapping of bands is minimal (1.5—1.8 eV). For the ZnS:Eu crystals we investigated we picked out the $h\nu = 1.7$ eV region, where there is practically no overlapping (on the edge of the E_2 band), but the photoluminescence intensity is still fairly high, allowing us to reliably measure the decay curves. An LGI-21 nitrogen laser ($h\nu = 3.68$ eV) with 10 ns pulses and pulse repetition of 10 GHz was used as the source of pulsed excitation. The photoluminescence was excited by radiation with an E-vector parallel ($E_\parallel$) and perpendicular ($E_\perp$) to the optical axis of the crystal. To do this, linearly polarized radiation was converted with a quarter-wave plate into circular polarization, then the necessary E-vector directions were selected out with a polarizer. The laser beam was focused into a square on the investigated part of the crystal with a cylindrical lens. The luminescence intensity $I(t)$ was measured over time intervals of 100 ns to 40 μs for two analyzer settings — parallel ($I_\parallel(t)$) and perpendicular ($I_\perp(t)$) to the optical axis of the crystal. Then the polarization kinetics are calculated as per Eq. (1). Since it turns out that the ZnS:Eu decay process is hyperbolic in nature, then the frequency of photoelectron pulses in the initial stages should be significantly higher then in later stages. With this in mind we constructed a time analyzer based on 20 CAMAC-standard binary counters and a specially-designed module — a pulse distributor (Fig. 8). The time analyzer cycle is started each time a pulse arrives from the clock pulse generator ($\nu = 10$ Hz) at the distributor: the laser is started and after a 100 ns delay the processed signals from the photodector begin to arrive at the first of the 20 counters. (This delay is necessary in order to quell the transient processes caused by pickup from the laser.) Each succeeding counter is gated at the instant the previous one is shut down and counts

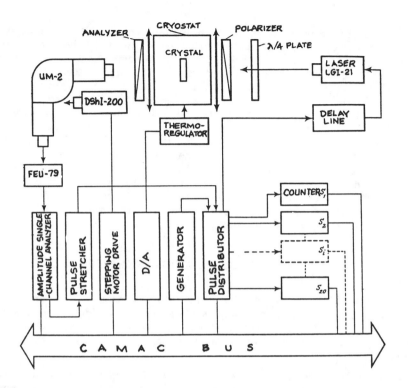

Fig. 8. Block Diagram of the Measurement Apparatus for Studying the Kinetics of Photoluminescence Polarization of Single Crystals in the Process of Afterglow

pulses over a time interval twice as long as the preceeding one. Thus the first one counts for 100 ns, the second for the next 200 ns, the third for 400 ns, and so on. After each cycle, information is transferred to a data array in computer memory and the counter resets. In this way we realized a 20-channel log-scale time analyzer on base 2. This turned out to be convenient, since each counter collected about the same number of pulses, and thanks to this the statistical measurement error of count rate obtained for each counter was about the same. The first counter was active for only a short time, however the rate of pulse arrival was high. At the last counter the pulses arrived slowly, but they were counted over a significantly longer time.

In the course of luminescence decay measurements a characteristic curve showing the reliability of the data stored in each counter was displayed on the graphics monitor screen every 100 s. The option of interrupting is then presented to the experimenter. Further processing of the data introduces a correction for the analyzer dead time (~50 ns) and time normalization (determined from the count rate of each counter), then it is output to the outboard equipment.

The results of the experiments turned out to be completely unexpected. As we already noted, according to our assumptions, if the second hypothesis is true (the first will be considered discarded) then the degree of polarization should fall off with the luminescence decay, as is usually observed in experiments; and if the third hypothesis is true the degree of polarization should remain constant. So it turned out that the polarization increased over the course of time! The corresponding results for one of the ZnS:Eu crystals is shown in Fig. 9. As can be seen, the luminescence polarization grows almost from zero and reaches a limiting value of $\sim$30% after $\tau \approx$50 μs. A similar picture is observed in other spectral regions and for different samples.

Table Degree of Polarization in ZnS:Eu Luminescence Bands for Various Moments of Time t

Maxima of the Band, in eV	Time Interval t, in μs	
	10 — 20	500 — 630
$E_1 = 1.78$	31	48
$E_2 = 1.96$	16	31
$E_3 = 2.22$	6.7	8.3

From these experiments it follows unequivocally that the preferred orientation of the radiators in the luminescence centers arises only at some stage of de-excitation as it is absent before excitation of the crystal. This at once explains the fact that the luminescence polarization is independent of the polarization of the pumping radiation: since the luminescence starts to become polarized only long after (on an atomic scale, that is) the action of the pumping radiation has stopped, it is clear that the reason for this polarization cannot be directly connected with a property of that pumping radiation.

The described results are relative to the $h\nu = 1.7$ eV region where, as we said, overlapping of bands is minimal. Direct study of polarization kinetics of other bands are difficult because of a high degree of overlap. However, another method can be utilized — decomposing the afterglow spectrum at different moments of time into its constituent elemental bands for two analyzer settings, and then knowing the contribution of each component of the spectrum, calculate the degree of polarization in each band. This operation was carried out for one of the ZnS:Eu crystals over two time intervals: 10—20 and 500—630 μs. The corresponding spectra are presented in Fig. 10 and the results of the degree of polarization calculation are presented in the above table.

As is evident from the table, the degree of polarization of the $E_1 - E_3$ bands grows with time, in spite of some quantitative differences. We were unable by this method to measure the kinetics of the E_4 band of this sample because the contribution of this component was small compared to measurement error in a number of cases. However, its kinetics were measured directly, since the short-wavelength end of the E_4 band is free from overlapping. It turns out that its polarization does not change during afterglow (Fig. 11), i.e. it acts like it is from a different sample compared to

127

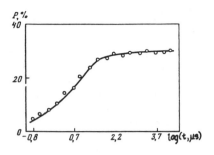

Fig. 9. Anomalous Kinetics of Luminescence Polarization of ZnS:Eu Single Crystals in the 1.7 eV Region for 10 ns Pulsed Laser Excitation ($h\nu = 3.68$ eV)

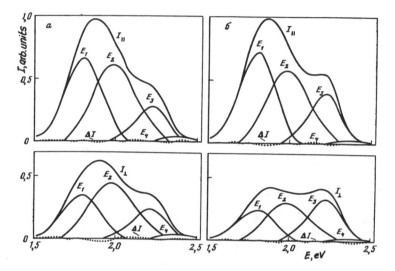

Fig. 10. Decomposition of the Afterglow Spectrum of ZnS:Eu Crystals Into Elemental Bands E_1-E_4 at Different Moments of Time ($a = 10-20$ μs, $b = 500-630$ μs)

$I_{\parallel}$ and $I_{\perp}$ are luminescence intensities at orthogonal analyzer settings; ΔI is the discrepancy curve of the decomposition

the E_1-E_3 bands, which increases the doubtfulness of its belonging to europium. We remark in passing that the discrepancy curve 6 on Fig. 10 lies practically on the abscissa, which attests to the fact that it changes little over the course of time, as well as to the accuracy of finding the elemental bands. However, the degree of polarization in afterglow still fails to reach 100%. The question of the reason for this depolarization remains open.

Not promoting any hypotheses on this score as of yet, we will consider what conclusions can be drawn from the experimental facts. As is evident from Fig. 9, the increase in the degree of polarization is noticeable even in the first half of a microsecond and, although it is completed after approximately 50 μs, polarization is maintained further for several microseconds. This means that there is at least enough time for the radiator orientation factor in the luminescence centers to oppose thermal disordering.

We will consider what processes will fit into this time frame. Since radiative transition is forbidden in Eu^{2+}, then the excited state lifetime does not exceed a few tens of nanoseconds, i.e. it comprises an insignificant fraction of this period of time. Consequently, this orientation factor acts on the luminescence center before its transition to an excited state, i.e. when it is found in an ionized state. Two cases are, generally speaking, possible. Either the luminescence center is still anisotropic just before ionization and only turns its axis because of the orienting factor, or it is isotropic before ionization and the anisotropy itself arises by the action of the orienting factor on the ionized center, and does not disappear over the lifetime of the center's excited state after recombination.

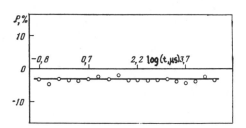

Fig. 11. Kinetics of Luminescence Polarization of the Short Wavelength E_4 Band of ZnS:Eu Crystal in the Region $h\nu = 2.24$ eV

The first case corresponds to an associative center. At room temperature the diffusion coefficient of impurities in zinc sulfide and its self-diffusion are so small that after a tenth of a microsecond only an insignificant part of the impurity ions or intrinsic defects have managed to cross over from one lattice site to another (or from one interstitial site to another). How small this fraction is can be seen from the following estimate. According to the data in [9], activating powdered zinc sulfide by the diffusion method with even a mobile impurity like copper requires firing for tens of minutes at temperatures higher than 300° C. The diameter of the powdered zinc sulfide luminophor granules is known to be on the order of 10 μm. Assuming for this estimate that copper can penetrate 10^4 interatomic distances (i.e. 3—4 μm) in 10^3 s (i.e. 17 min), and it can traverse one interatomic distance in a period of time 8 orders of magnitude less, i.e. 10^{-5}s. But this is at 300° C, at room temperature the movement will be slower, corresponding to the diffusion activation energy. Different authors have put forward different values for this quantity, but none are less than 0.4 eV. Taking that for the diffusion activation energy, we obtain that at room temperature it will proceed three orders of magnitude slower than at 300° C. (In point of fact, it is probably even slower than that.) This means that one jump takes no less than 10^{-2}s. Consequently, after 1 μs only 10^{-2}% of the luminescence centers can have been turned, which in no way could have caused the 5% increase in the degree of polarization observed in experiment. The time for a jump for a less mobile impurity such as europium can only be greater, which even further decreases the possible degree of radiator orientation. It follows from this that the luminescence centers formed by Eu^{2+} ions in ZnS:Eu are in all probability not associative.

This conclusion was also reached in [10]. Henceforth we should use the reasoning expressed there, and thus restate what the basis is for this conclusion in two sentences. It was demonstrated in that article that a part of the Eu^{2+} ions have an EPR spectrum corresponding to cubic center symmetry, and part are found in a field with a distinct axis (parallel to the C axis of the crystal), although this symmetry is transformed to cubic during plastic deformation of the crystal. Only a center formed by identical ions may have cubic symmetry, and an associative center may have axial symmetry, but its distinct axis does not disappear during plastic deformation and if it is turned, that is not observed experimentally.

Thus we are dealing with centers formed by identical ions of Eu^{2+} without other impurities nor point lattice defects nearby. But the symmetry of the field about the ion can change if the ion itself is displaced from the center of the tetrahedron it is supposed to be in. During ionization of the luminescence center this very ion changes its charge (becoming Eu^{3+}) which unavoidably will affect the force of the crystal field on it. If this field has an axial component, then it will act to displace the europium ion along this axis. As is known, the ZnS hexagonal interlayer has such an axis directed along the C axis of the crystal. Therefore the displacement will be along the C axis (Fig. 12), i.e. along a line joining the center of the tetrahedron and one of its vertices corresponding to the direction of the E-vector of the emitted radiation.

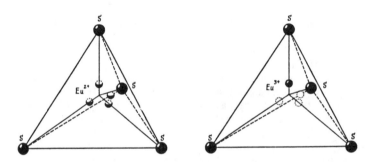

Fig. 12. Eu Ions in the ZnS Lattice Before (a) and After (b) Ionization

However, the question arises as to why this displacement happens so slowly. After all, the new location of the potential energy minimum comes into being instantly after the ionization of the center, and the Eu^{3+} ions (arising from the Eu^{2+}) should transit to there in something on the order of the period of oscillation of high-energy phonons, i.e. in 10^{-12} to 10^{-11}s. This is at least five orders of magnitude more rapid than that experimentally observed. Apparently the delay comes from the fact that the transition to the new europium ion equilibrium position has to overcome some sort of potential barrier.

DEPENDENCE OF POLARIZATION KINETICS ON TEMPERATURE

Before we clarify the nature of this potential barrier we must convince ourselves experimentally of its existence. With this in mind we studied the effect of temperature on the rate of change in the degree of luminescence polarization during afterglow. The experiment was conducted with the same set-up except that the crystal was placed in a metal circulating cryostat cooled by pumping a refrigerant through it. The cryostat was made up of three basic parts: the housing, the heat exchanger and the cavity for the sample being investigated. The housing and cavity had quartz windows. Thermal insulation was provided by a vacuum jacket or sleeve and copper radiating baffles. The cavity was filled with an inert gas for more even cooling of the crystal and to prevent condensation of moisture on its surface and on the quartz windows. The frigid vapors from liquid nitrogen entered the heat exchanger directly from a Dewar flask in which the pressure was a little more than atmospheric. A nichrome wire wound evenly around the heat exchanger acted as a heating element. The temperature in the cavity was determined by the pump rate of nitrogen vapor and the thermal power evolved at the heating element. Heat exchanger and sample temperature probes were standard KD514 semiconductor diodes. The voltage drop over the forward-biased p-n junction was measured with a digital voltmeter and this was cross-referenced with a calibration chart to determine the temperature. A coarse indication of cryostat temperature was given by the pump rate of nitrogen vapor. Temperature stability about a point was accomplished automatically by a signal from a diode located on the heat exchanger. If the temperature of the heat exchanger dropped below a given value then the current to the heating element was increased, and vice versa. Thanks to thermal inertia the amplitude of temperature fluctuations in the cavity was significantly less than at the heat exchanger where the probe for the automatic device was mounted. Crystal temperature instability did not exceed 0.5° C.

The results of these measurements are presented in Fig. 13 on a log-log scale. As is apparent, as temperature increases the increase in polarization is dramatically slowed, and the increasing polarization curve becomes

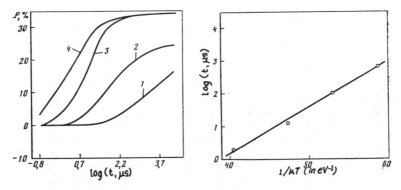

Fig. 13. Kinetics of Luminescence Polarization of ZnS:Eu During Afterglow Processes at Various Temperatures in the Spectral Region $h\nu = 1.7$ eV

1 - T=200 K; 2 - T=222 K; 3 - T=248 K; 4 - T=292 K

Fig. 14. Temperature Dependence of the Time $t_{1/2}$ Needed to Reach Half Maximum of the Degree of Luminescence Polarization in ZnS:Eu

deformed. This prevents direct measurement of the process activation energy from displacements of the curve. For a rough estimate of the activation energy we have to use the time $t_{1/2}$ as a characteristic position on the time axis of the curve describing the growth in polarization; $t_{1/2}$ is the time necessary for the degree of polarization to attain half its maximum value. Fig. 14 depicts the dependence of $t_{1/2}$ on inverse temperature (to be precise, on $1/(kT)$, where kT is in eV). As is evident, the points fall fairly well on a straight line, as they should if the barrier has a definite height. The activation energy found from the slope of this straight line was equal to 0.37 eV. At room temperatures a barrier of this height would slow down the process by about six orders of magnitude. This means that if europium ions could transit to the new equilibrium position in $1-10$ ps without the barrier, then the time for the transition will be $1-10$ μs due to the necessity of overcoming this barrier, and this conforms with $t_{1/2}$ at room temperature.

As to its origin, this might be a barrier between Jahn-Teller potential energy minima for europium ions in a tetrahedron formed by sulfur ions. For zinc ions in the cubic phase all these minima (potential wells) are practically equivalent, since according to crystallographic data the center of gravity of this ion is in the center of the tetrahedron. It is possible to say this as regards the Eu^{2+} ion because it, like the zinc ion, is divalent. If the Jahn-Teller wells are displaced from the center of the tetrahedron along a third-order axis, then those wells found on the C axis will be the deepest. Thus the Eu^{3+} ions may be delayed in them longer than in the other three. Concerning europium ions located inside the hexagonal interlayer, as was shown in [2], they are slightly displaced from the center of the tetrahedron even in an unexcited crystal, i.e. when they are still in the divalent state. This leads to the fact that Eu ions are delayed in Jahn-Teller wells located along the C axis even in the divalent state. The transition of Eu^{2+} ions to the ionized Eu^{3+} state during pumping inside the hexagonal interlayer can but augment this effect.

A gradual transition from an equilibrium distribution of Eu^{3+} ions among the four wells which exists immediately after ionization of the luminescence centers in cubic phase regions, or from the near-equilibrium distribution in hexagonal interlayers, to a distribution with a predominant population of one of them located on the C axis must cause an increase in the degree of polarization after cessation of pumping, which is indeed observed in experiment. As one should expect, this is especially strongly manifested in the E_1 and E_2 bands which, according to [10], are ascribed to Eu ions in the cubic phase of ZnS crystals. As follows from Table 1, the degree of their polarization increases over time by a factor of about $1.5-2$, at the same time as that for the E_3 band increases by only 20%. According to [10], the E_3 band belongs to europium ions in the hexagonal interlayers. An absence of variation in the degree of polarization of the E_4 band (Fig. 11) makes us think that in fact it belongs to luminescence centers of some other origin.

However, the remaining wells are not completely depopulated. The situation is that distance between them and the deeper wells in the same tetrahedron is no more than tenths of an ångstrom. The electric field strength of the hexagonal interlayer crystal field is on the order of about

10^7 V/cm, which over a distance of tenths of an ångstrom yields an energy only in the hundreds of eV. This is on the same order as the difference in depths of the Jahn-Teller wells caused by the crystal field. At room temperature this corresponds to only a few hundred percent difference in the population of the wells. The highest degree of polarization observed at room temperature (32%, as per Fig. 9) is for a situation in which 49% of the radiators are found in the deepest well and the rest (51%) are distributed equally among the other three. This means that the population of the deepest well is a little more than three times higher than the population of any one of the shallower wells. Thus both estimates agree well with each other.

It would seem that if the lack of 100% polarization is caused by thermal disordering of the radiators, then the ultimate degree of polarization actually attained should go up as the temperature is lowered. This was observed experimentally only down to about 250 K. Upon further cooling the ultimate degree of polarization decreased. This is easily explained by recombination interactions between the luminescence centers, i.e. exchange of holes between ionized and non-ionized centers. This exchange also results in depolarization, because a hole coming from a Eu^{3+} ion that already occupies the deepest well will, three times out of four, run into a Eu^{2+} ion in one of the other three wells, which after the ion captures the hole will turn out to be not the deepest. If the recombination interaction requires a lower activation energy than the migration of a Eu^{3+} ion from one Jahn-Teller well to another, then as the temperature is lowered the recombination mechanism will be moderated less and less and, consequently, its role will be increased. For a quantitative check of this proposition we need independent experiments to determine the depth of the Eu^{3+} hole level which is as of now unknown. Thus we note that the drop in the degree of polarization observed experimentally can be explained if the depth of the Eu^{3+} hole level is $0.05-0.08$ eV less than the barrier height between Jahn-Teller wells (as viewed from the shallower of the wells, which was also determined from the experiment). In this way we obtained a Eu^{3+} level depth of 0.3 eV. This is an entirely reasonable value.

This picture is in agreement with the mentioned experiments on electron paramagnetic resonance in europium centers. In fact, moreso than the divalent ions, the trivalents interact strongly with the axial as well as with other components of the crystal field. Therefore the potential barrier between Jahn-Teller wells will be higher for them than for the divalents. If we take for a first approximation that the height of this barrier is proportional to the charge, then we obtain a barrier to Eu^{2+} ions 0.12 eV lower than to Eu^{3+}. At room temperature this corresponds to an increase in the barrier tunneling rate by about two orders of magnitude. As we already stated, a single tunneling requires about 1 μs ($t_{1/2}$ at room temperature); consequently, Eu^{2+} requires about 10 ns. In EPR experiments the resonator had a Q of several thousand, i.e. averaging the readings for that number of periods of the electromagnetic field. Consequently for a frequency of that field of 10 GHz the averaging goes on for several hundred nanoseconds, i.e. a much longer time than an Eu^{2+} stays in any one well. This leads to the fact that, from the EPR standpoint, the Eu^{2+} would always be found in the center of the tetrahedron. Only those ions located inside the hexagonal interlayer interact so strongly with the crystal field that their EPR feels the lowering of the symmetry of the surrounding field. Eu^{3+} ions, on the contrary, feel the hexagonal interlayer field even when they are in cubic-packing layers, but they yield no EPR. We observe their arrangement only indirectly from the radiation formed by them during recombination of those Eu^{2+} ions which, over the lifetime of the excited state, cannot leave the sites that the europium fell into when it was still a trivalent ion.

POLARIZATION KINETICS OF ZnS:Tm

The suppositions about the reason for the anomalous kinetics of luminescence polarization in ZnS:Eu were checked indirectly by yet another independent method. If the situation is indeed such that trivalent europium ions feel the crystal field axial component significantly stronger than divalents, then the same must hold for other rare-earth elements if they are doped into zinc sulfide in divalent form. However, all of them are doped in their trivalent form. Therefore the radiators in them must still be oriented before excitation of luminescence. As is known, during recombination luminescence mechanisms there arises ionization of the luminescence centers, i.e. centers which have changed their charge. In the case of trivalent rare-earth elements the ionization of luminescence centers consists of having a rare-earth ion capture an electron and change its charge from $+3q$ to $+2q$ (where q is the absolute value of the charge of an electron). Recombination consists of hole capture by such centers, as a result of which the rare-earth ion transits to an excited state and its charge becomes $+3q$. In this state it also luminesces. For as long as the ion is in the divalent state it may easily transit to another Jahn-Teller well, and after recombination it may not and cannot return. Therefore the degree of polarization should be greatest right

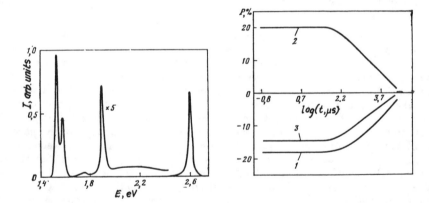

Fig. 15.   Luminescence Spectrum of ZnS:Tm Single Crystals at Room Temperature and Steady-State Pumping

Fig. 16.   Kinetics of Luminescence Polarization of ZnS:Tm at 300 K for Bands at $h\nu = 1.54$ (1); $h\nu = 1.9$ (2); and $h\nu = 2.6$ eV (3)

after excitation by a laser pulse and will gradually fall off over time. A value for this effect is, however, unknown since radiative transitions in trivalent ions of rare-earth elements are forbidden, and thus the lifetime of the excited state may, generally speaking, exceed the time needed to overcome the barrier between Jahn-Teller potential wells. Therefore the possibility is not excluded that, before it is de-excited, a trivalent ion can transit to a deep well and orient its radiator along the C axis.

For the experiments thulium was taken as the activator, which already at room temperature yields several narrow luminescence bands in zinc sulfide (Fig. 15). Preliminary investigation demonstrated that for constant pumping, the most intense luminescence band of ZnS:Tm is strongly polarized. The bands with maxima at $h\nu = 1.54$ and 2.6 eV had an E-vector directed predominantly along the C axis, i.e. in contradistinction to ZnS:Eu the polarization is negative. This difference is easily explained by the forbiddenness of radiative transitions in Tm^{3+} in the dipole approximation. In terms of the value of the transition probability, the allowed electric dipole transition follows the magnetic-dipole transition, which gives the observed azimuth of polarization of each band. If the forbiddenness of the dipole transition were not so strict the polarization would remain positive. Apparently, this also takes place for transitions responsible for the band with $h\nu = 1.9$ eV. As we anticipated, in contradistinction to Eu^{2+} bands, all these bands are actually more strongly polarized at the onset of afterglow and thereupon their degree of polarization decays to practically zero (Fig. 16). The time for decay to half, despite expectations, turned out be around 2 ms, i.e. about three orders of magnitude greater than the time for the increase in luminescence polarization of ZnS:Eu.

First of all the question comes up, how real is this decrease in polarization, might not the convergence of readings for two analyzer settings be simply tied in with an overall decrease in intensity and an increase in the role of phonon pulses as a result, for which the correction cannot possibly be complete? For a check on this same equipment we investigated the time-dependent behavior of the luminescence polarization of ruby which, as is known, is distinguished by a high degree of sponataneous polarization. It turned out that, over all investigated time intervals the degree of polarization of its luminescence at $h\nu = 1.78$ eV was maintained constant at $74-75\%$. The luminescence intensity decreased by more than two orders of magnitude over this time. In order to include the final interval of ZnS:Tm luminescence intensities which are of greatest interest to us, the experiment was repeated with the ruby luminescence attenuated by about an order of magnitude before the monochromator entrance. The results came out the same. This demonstrates that phonon pulses have been correctly taken into account and we need to search for a physical reason for why the degree of luminescence polarization decreases so slowly.

Understanding the reason for this is possible, apparently, if we compare the luminescence decay behavior of ZnS:Eu and ZnS:Tm crystals. In Fig. 17 is depicted on a log-log scale the afterglow behavior of these crystals over the same bands depicted in Figs. 9 and 16 for the time-dependent polarization behavior. The sum of readings at one or another analyzer setting was taken as a measure of the luminescence intensity. As is evident from this Figure, only in ZnS:Eu does the luminescence decay follow a hyperbolic path

135

with gradual increase in the exponent of the hyperbole as is usual for recombination radiation mechanisms. The ZnS:Tm bands decay according to a more complicated law. As such, the $h\nu_{max}=1.9$ eV band almost does not change in intensity over the first 10 μs and only after 100 μs begins to fall off by something approaching a second-degree hyperbole. The $h\nu_{max}=2.6$ eV band behaves similarly, the only difference being that around 4 μs it has a barely noticeable maximum. The $h\nu_{max}=1.54$ eV band also has a maximum (at 15 μs) which is much more clearly expressed.

The existence of such a maximum unequivocally shows that the crystal has some sort of "reservoir" or intermediate level that the energy gets to after the action of laser pulse excitation, where it is maintained for tens of microseconds and only then transferred to the luminescence centers. Generally speaking, there are two types of "reservoirs". First, it may be simply a highly excited state of the center to which transition is made during recombination. The second type of "reservoir" is a trap for electrons and holes which they fall into during or shortly after the action of the excitation pulses.

We will look at the first possibility. If the center makes a transition during recombination chiefly to this high level and lower-lying levels remain underpopulated, then for some time after the excitation pulse its population will increase. The height of the population maximum, and this means too the luminescence intensity maximum, depends on the initial population of these two levels as well as the ratio of their lifetimes and the probability of transitions between them. In particular, it may turn out that there is no maximum but the luminescence intensity is maintained almost constant for a long period of time.

This picture is very similar to that depicted in Fig. 17, however, it contradicts the results of polarization measurements illustrated in Fig. 16. Actually, if there were a delay in de-excitation caused by a small probability of intra-center transitions, then the lifetime of the Tm^{3+} excited state participating in these transitions would be no less than $10-20$ μs, since only at that time does the afterglow intensity begin to drop off. But soon after the recombination processes are over and the Tm ions return to their trivalent state, they will again try to occupy the deepest Jahn-Teller well. If they succeed, they will still be excited since, as was already stated, to do this takes only about a microsecond, i.e. 10 times less than its excited state lifetime. And this means that the degree of luminescence polarization will remain constant the entire time. Although in the experiment the depolarization was retarded, it occured all the same. This is why the first possibility has to be rejected.

It remains for us to look at the second possibility — shallow electron and hole traps. It is well known that nonequilibrium charge carriers arising from short wavelength light pulses are at first distributed among the centers not according to the depth of their levels, but according to the concentrations and effective capture cross-sections and only later do they gradually make the transition from shallow levels to deeper ones. Independent of the depth of the shallow levels, such processes may go on for some time or may occur relatively quickly, e.g. in microseconds. If these deeper levels have non-ionized luminescence centers, then in the process of afterglow they will be ionized as a result of capture of charge carriers of the given sign.

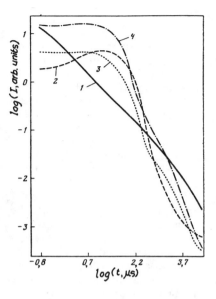

Fig. 17. Afterglow Behavior in ZnS:Eu Single Crystals for $h\nu = 1.9$ eV (1) and ZnS:Tm for $h\nu = 1.54$ eV (2); $h\nu = 1.9$ eV (3); and $h\nu = 2.6$ eV (4)

This is why there occurs in afterglow an increase in ionized luminescence centers, not a decrease. Which of these processes is predominant depends on the initial concentration of ionized centers and charge carriers in shallow traps and, of course, on their parameters as well as on the parameters for traps of charge carriers of the opposite sign. In our case the ionized centers, as we already stated, are Tm^{2+} and the non-ionized are Tm^{3+}. The maximum concentration of Tm^{2+} in afterglow will happen if at first the photoelectrons are captured primarily by the shallower traps and only then make the transition to Tm^{3+}, turning it into Tm^{2+}. But this is still not a sufficient condition for the appearance of an intensity maximum. For this we still require that the concentration of free holes does not decrease too rapidly, i.e. that the rate of its decrease (if this occurs) does not exceed the rate of increase in concentration of the ionized luminescence centers. If these rates over the course of time become equal, then the afterglow intensity will become constant. As is apparent, this has the same variations in afterglow behavior as in the first case.

The polarization of the luminescence in this case still will not be constant. Actually, the time interval between the instant that Tm^{3+} captures an electron and turns into Tm^{2+}, and the instant a hole recombines with Tm^{2+} returning it to an excited Tm^{3+} ion, depends on the hole concentration. If this time span is less than a microsecond immediately after pumping, then the Tm^{2+} ions cannot leave the Jahn-Teller wells in which they were generated from Tm^{3+}. Therefore the luminescence will be polarized even if the Tm^{3+} excited state lifetime is short. But as the supply of nonequilibrium holes is used up, their concentration in the valence band is lowered and Tm^{2+} ions have to wait even longer for holes. Sooner or later this "waiting" time will become longer than the time needed to overcome the potential barrier between two Jahn-Teller wells. If after the excited state lifetime a Tm^{3+} ion has not succeeded in returning to a deep well, then starting at that instant the degree of luminescence polarization begins to decline. When the "waiting" time for holes significantly exceeds the time for tunneling of a Tm^{3+} ion from one well to another, then the degree of polarization becomes practically zero.

137

CONCLUSION

We return now to the question posed at the start of this article about the reason for the degree of ZnS:Eu (and likewise ZnS:Tm) luminescence polarization being different from 100% during steady-state pumping. The disorienting of radiators in luminescence centers connected with the Jahn-Teller effect discussed above doubtless plays some role in this. However, even in those cases where one might think that all radiators are completely oriented, i.e. at the end of afterglow in ZnS:Eu and at the start of afterglow in ZnS:Tm, the degree of polarization still does not reach 100%. In the case of ZnS:Eu, there is only one conclusion left, that the precession of radiators does after all play the fundamental role in depolarization of the luminescence, as was supposed in [11]. However, differently from what was supposed in that article, the precession axis itself may take on various directions and the degree of orientation of the radiator precession axes in the ensemble of luminescence centers is determined by the Jahn-Teller effect. In the case of ZnS:Tm these two circumstances are augmented by yet one more. As follows from the previous article in this collection which is also devoted to ZnS:Tm, the bands we investigated were formed by transitions between several sublevels of different terms. Each transition gives radiation with its own degree of polarization differing from the others in absolute value (e.g., due to radiator precession) and sometimes also in sign. In the experiment we measured only the averaged value of the degree of polarization which was invariably less than 100%. As concerns the orientation of the radiators, it may, as mentioned in that article, be almost total at helium temperatures, since the degree of polarization of a number of lines at that temperature were significantly higher than 50%.

This article is the first detailed study of the influence of the Jahn-Teller effect on luminescence properties of rare-earth crystal phosphors. Naturally, more questions were brought up than were solved. Further research obviously should follow, studying the kinetics of luminescence and its polarization at various temperatures and pumping intensities over as wide a range as possible of these conditions.

In conclusion, the authors would like to express their appreciation to N. A. Gorbacheva, A. G. Glyadelkina, A. V. Lavrov and L. M. Tsyganova for the crystals which they kindly loaned for our research.

BIBLIOGRAPHY

1. Arkhangel'skiy, G. Ye., Bukke, Ye. Ye., Voznesenskaya, T. I. et al, "On the Breakdown of Isomorphism During Substitution of Ions in the ZnS Base Lattice for Impurity Ions" in the book: "Voprosy izomorfizma i genezisa mineral'nykh individov i kompleksov", Elista, Kalmyts. Univ., 1977, part 2 pp 311-316

2. Arkhangel'skiy, G. Ye., Bukke, Ye. Ye., Voznesenskaya, T. I. et al, "Visualizing the Structure of Defects in ZnS-Type Crystals by the

Anthrazyne Decoration Method" in the book: "Metody vizualizatsii izobrazheniy" [Image Visualization Methods], Moscow, Nauka, 1981 (TRUDY FIAN Vol 129)

3. Fok, M. V., "Decomposition of Complicated Spectra Into Individual Bands With the Generalized Alentsev Method" in the book: "Lyuminestsentsiya i nelineynaya optika" [Luminescence and Nonlinear Optics], Moscow, Nauka, 1972 (TRUDY FIAN Vol 59)

4. Fok, M. V. and Yakunina, N. A., "Effect of the Crystal Lattice Field on the Luminescence of Europium Ions in ZnS Crystals", KRAT. SOOBSHCH. PO FIZIKE FIAN No 6 pp 9-14, 1980

5. Averyushkin, A. S., Blazhenkov, V. V., Vitukhovskiy, A. G. et al, "Automating Optical Experiments With a Two-Level Computer System", PREPRINT FIAN No 118, Moscow, 1982

6. Podol'skiy, L. I., "Sistema QUASIC dlya programmirovaniya na mini-EVM" [The QUASIC System for Programming Mini-Computers], Pushchino, NTsBI; NIVTs, 1980

7. Tiberg, Ya. E. and Paulauskas, V. N., "Single-Electron Pulse Selector for a Photon Counter System", PTE No 5 pp 183-186, 1980

8. Ovchinnikov, A. V., "CAMAC-Standard Color Graphic Monitor for Displaying Images of Time-Resolution Spectra", PREPRINT FIAN No 193, Moscow, 1983, 25 pp

9. Ril', N., "Lyuminestsentsiya" [Luminescence], Moscow/Leningrad, OGIZ, 1946, pp 105-114

10. Arkhangel'skiy, G. Ye., Grigor'yev, N. N., Fok, M. V. and Yakunina, N. A., "Effect of Plastic Deformation on the Luminescence and Electron Paramagnetic Resonance of ZnS:Eu Crystals" in the book: "Lyuminestsentsiya i anizotropiya kristallov sul'fida tsinka" [Luminescence and Anisotropy in Zinc Sulfide Crystals], Moscow, Nauka, 1985, pp 44-102 (TRUDY FIAN Vol 164)

11. Grigor'yev, N. N., Bukke, Ye. Ye. and Fok, M. V., "Application of Polarization Diagrams for Studying Single-Axis Crystals" in the book: "Tsentry lyuminestsentsii v kristallakh" [Luminescence Centers in Crystals], Moscow, Nauka, 1974, pp 108-144 (TRUDY FIAN Vol 79)

The Mechanism of Electrolytic Activation by Rare - Earth and Other Elements in Single Crystals of Zinc Sulfide

G.Ye. Arkhangel'skiy, Ye.Ye. Bukke, T.I. Voznesenskaya
N.N. Grigor'yev and M.V. Fok

Abstract: It is demonstrated that ions of rare-earth elements, as well as copper and silver, can be doped into zinc sulfide crystals in 30 minutes under an electric field of 200 V/cm at 600° C if this field is oriented perpendicular to the hexagonal axis of the crystal. This method for the rare-earth elements, augmented with subsequent half-hourly firings at 1200° C without the field, allows us to activate an entire crystal some tens of cubic millimeters in size many times faster than by diffusion at the same or even higher temperature. It was discovered that half-hourly firings at 600° C markedly decreases the number of hexagonal phases in the crystal, while firings of the same duration at 1200° C followed by rapid cooling causes an increase.

Difficulties often arise in the growing of activated single crystals in connection with the fact that the conditions for growing a perfect host crystal do not coincide with the conditions for successfully introducing the activating impurities. These difficulties are increased if one must grow a crystal with two impurities, especially when they differ greatly in fugacity. In the preparation of p-n junctions, usually one of the impurities is added when the crystal is being grown, and the other is diffused into the already-grown crystal. However, this method only works for relatively small diffusion depths. For example, according to the diffusion equation, going from activation depths in microns (sufficient for making p-n junctions) to depths in the millimeter range requires a million-fold increase in the diffusion time! It is, of course, possible to conduct the process at elevated temperatures to increase the diffusivity of the introduced impurity, but this on more than one occasion has managed only to exceed the limiting thermal stability of the crystal.

One method of introducing impurities which is free of these defects is the electrolytic method, in which the doping ions move in a directed fashion under the influence of an electric field. The velocity of their translation is proportional to the electric field strength and does not decrease with time, as happens in the diffusion method. Therefore, given a sufficiently strong electric field, the time necessary for doping a crystal on the order of millimeters in size turns out to be several orders of magnitude less than that with the diffusion method. However, the method has its own difficulties connected with the fact that, in compliance with the Einstein relation, the mobility of the carrier is proportional to its diffusion coefficient. Therefore this method also requires elevated temperatures. However, the electron current also grows during heating. If it grows faster than the ionic current, it may turn out that thermal breakdown occurs before the drift velocity of the impurities reaches the required value. Thus, finding the best conditions for introducing impurities is a problem which must be solved on a case-by-case basis for each crystal-activator pair.

If the charge or size of an impurity ion introduced into a crystal lattice site differs greatly from that of the host ion it replaced, then this difference must be compensated, as is known, with the help of the intrinsic defect lattice or ions of the other impurity. Rare-earth ions introduced into zinc sulfide (zincblende) are no exception to this rule. The majority of the introduced ions are trivalent, for which charge compensation is obviously necessary. Size compensation is not so important in this case, since the ionic radius RE^{3+} exceeds that given by the Goldschmidt rule for the limiting size of easily-introduced impurities by only a small amount. (The ionic radius RE^{3+} is 17% larger than the ionic radius of Zn^{2+} instead of the 15% limit according to the Goldschmidt rule.) Europium is introduced in its divalent state, necessitating no charge compensation, but on the other hand necessitating size compensation, in that the ionic radius of Eu^{2+} is significantly larger than the radius of Zn^{2+}. Thus although we have suceeded in doping zinc sulfide with rare-earth elements during crystal growth, due to the necessity for charge and size compensation of the impurity ions several defects arise, which results in an imperfect crystal in the crystallographic sense. The situation is especially bad during simultaneous doping with two activators, such as a rare-earth element and copper, which is required to make the crystal elecroluminescent. In this regard the electrolytic method only seems attractive for doping with one of these impurities. First, however, the mechanism of electrolytic implantation of impurities should be understood and it should be established whether the centers thus formed are the same as those which occur when the same impurity is doped in during crystal growth.

ACTIVATION METHOD

The electrolytic method of activating zinc sulfide was first utilized in 1961 in [1]. However, the information of interest to us about the mechanism of electrolytic doping was only obtained in part from the much later articles [2] and [3], which confirmed the fact of directed motion in the electric field by impurity ions of both charge signs in zinc sulfide. Tests were conducted on pure, previously deoxidized zinc sulfide powder, which was placed in a small quartz tube between two graphite electrodes. An interlayer of the crystalline phosphor ZnS—Cu, Cl with green luminesence was put in the middle of the unactivated ZnS. Thereupon the quartz tube with the sample was placed inside an larger quartz tube, which was then sealed on one end and the air inside replaced with argon. The entire apparatus was then placed in an oven and fired at $T=1060°$ C for 12 to 15 h. During this time an electric current on the order of 2 to 8 mA was fed through the sample. The sample which was fired and subjected to the current became photoluminescent. Small "belts" of orange glow 2 mm thick were observed near the cathode. Further from the cathode this glow changed to a green. The ZnS—Cu, Cl interlayer itself became practically nonluminescent at this point. A blue glow appeared near the anode in those cases where the initial crystalline phosphor in the interlayer region was prepared from a NaCl melt.

These changes can be explained by proceeding from the fact that there is an ionic as well as an electronic current caused by the applied voltage, so that the copper ions are translated toward the cathode, and the chlorine ions toward the anode. This fact, that copper moves toward the cathode, shows that it has a positive charge with respect to the ZnS crystal lattice. A Cu^{2+} ion occupying a cation site has the same charge as the zinc ion for which it substituted, i.e. it is neutral with respect to the ZnS lattice. In this circumstance, it could only be displaced by such fields as would cause Zn ions to drift, i.e. would result in the disassociation of the zinc sulfide itself. The fields used in these tests were much less than that. The second possible state for copper in a cation site could be as Cu^+, which corresponds to a negative charge on copper with respect to the lattice. In such a state the copper ions would have to be displaced toward the cathode along with the chlorine ions. Interstitial copper alone has a positive charge with respect to the ZnS lattice, and that is precisely what is travelling toward the cathode. There is reason to believe that other impurities might be introduced into the crystal in this way, since less energy would be required for activation than for migration along vacancies.

Therefore in this article we employ a method for activating crystals which is very similar to that already successfully applied to powders. However, it we had to perform the tests numerous times before we managed to obtain sufficiently repeatable results. The chief difficulty was in making a contact with the crystal being activated which would ensure the ionic as well as the electronic components of the current. The device used for this —an electrolyzer— is depicted in Fig. 1. Wafers knocked out of bulk single-crystal ZnS (1) roughly $5\times5\times1$ mm in size in quantities of four were pressed onto a common graphite electrode (4) on a quartz stand (5) with contact springs (2) and graphite current leads (3). All of this was placed inside a quartz tube (6) with a sealed end and outside diameter of about 4 cm. Desiccating argon was admitted to the system through a central tube (7). A layer of activating impurities was applied to the side of the little crystals opposite the common electrode. For successful activator doping at relatively low electrolysis temperatures, the composition of this layer should ensure relatively easy dissociation of the salts containing the activator. During

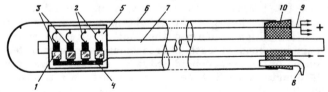

Fig. 1. Electrolyzer

1-Wafers of single crystal ZnS; 2-nichrome contact springs; 3-graphite current leads 4-common graphite electrode, 5-quartz stand, 6-quartz tube; 7-quartz tube for injecting argon; 8-tube for escape of gas; 9-contact leads, 10-rubber stopper

copper doping, thin layers of $CuSO_4$ and $CuCl_2$ deposited from solutions were used with equal success, as well as thin layers of Cu_2O powder. Silver doping occurred during deposition of a layer from a solution of $AgNO_3$. The choice of a dissociating salt for doping with rare-earth elements is a difficult one but, as the tests showed, it is possible to use powders of rare-earth oxides or sulfides. The small degree of dissociation of these compounds, with their high melting points and consequently low rate of ion generation at the electrode, was compensated for by using large quantities of the materials, hundreds and thousands of times more than required. A binder (sodium silicate solution) was used to give the activating layer mechanical stability, and a very small quantity of Aquadag was added to the activating layer to improve electrical conductivity. A layer of pure Aquadag was deposited on the surface of the activating impurity layer 0.1 to 0.2 mm thick after drying and heating to 250° C. A similar layer of Aquadag was deposited on the side of the crystal facing the common electrode.

The electrolyzer was placed in the same half of the cold oven where the sample was located. Thereupon the oven was slowly heated to a temperature of 600 to 650° C in a constant stream of desiccating argon. The entire heating process took about three hours, with a hold for a half an hour in the 350-400° C region where, as is well-known, intensive outgassing of the graphite occurs (part of this is the Aquadag). An electrolysis temperature of 600° C was selected because, on the one hand, it is high enough for relatively rapid electrolytic doping, and on the other hand, it is low enough to allow the use of this simple apparatus and technology for doping.

In addition to Cu and Ag, we tested Sm, Dy, Tb, Tm and Eu as activators. These impurities were doped into the crystal under the influence of an electric field strength of roughly 200 V/cm. The current density (chiefly electronic current) at that time was $(1—4) \cdot 10^{-4}$ A/mm^2. Jumping ahead for a moment, we remark that under these conditions in the crystal wafers all the aforementioned impurities were introduced and evenly distributed throughout the length of the crystal after an hour or even a half hour.

THE RAPID DOPING STAGE

We turn our attention now to two subjects: first of all, the extraordinarily fast drift speed of the impurity ions and, secondly, the anisotropy of the doping. It turns out that high-speed drift of impurity ions occurs only for certain crystal orientations. Electrolysis is facilitated when the opposing electrodes are located on the crystal planes containing the optical axis C, i.e. when the current flows in a direction nearly perpendicular to the C axis. Fig. 2 gives an example of such an orientation of the crystal planes. Practically no electrolysis takes place in the direction along the C axis. In the experiment we observed drift speeds more than 1000 times higher than that usually observed at the same temperature in diffusion penetration. For example, in [4] it was demonstrated that silver ions could be distributed in zinc sulfide by the diffusion method to a depth of tenths of a micron, under analogous conditions. The electric field we used, on the order of 200 V/cm, did not substantially lower the energy barrier to transitions of ions from one interstitial site to another. At the same time, and recalling the aforementioned almost total absence of doping electrolysis when the electric

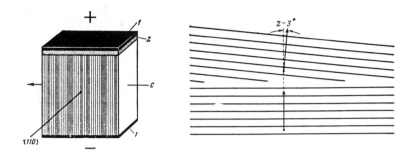

Fig. 2. Orientation of the Crystal Relative to the Optical Axis During Electrolytic Activation. The vertical bands are regions of the crystal differing in birefigence; 1- Aquadag electrodes; 2- activating impurity layer

Fig. 3. Schematic Drawing of an Edge Dislocation in ZnS Crystal. The direction of the lines corresponds to the packing planes of the atomic lattice. The long arrows indicate the direction of the optical axis in the strata.

field is directed along the C axis, this indicates that there is insufficient thermal energy for overcoming these barriers under the conditions in our experiment. We remind that the experiment described in this article was conducted at temperatures 400 to 500° C lower than in [2,3].

We must look for an explanation of the observed drift speed anisotropy to the presence of singularities in crystalline ZnS associated with its structural characteristics — polytypicism and the existence of a large number of plane defects — and flaws in the sequence of the densely-packed atomic layers. This leads to the boxy appearance of the crystal, and to the creation of relatively large and rather long channels possibly formed by dislocation planes, along which the impurity atoms are able to move much more freely than along the interstitial lattice.

The existence of plane defects in the crystals used in our experiment was revealed in the form of colored interference bands observed through crossed polaroids under a microscope, and shown in Fig. 2. As is known, bands of the same color identify a small region of the crystal characterized by a nearly constant birefringence. The birefringence is itself determined by the concentration of hexagonal layers in the lattice, which is primarily constructed with a cubic basis for layer packing . Neighboring crystal regions having a different color in the interference pattern are thus characterized by a different average concentration of hexagonal layers.

Earlier [5] we discovered that these regions are wedge-like in shape and that the direction of the optical axis may change $\pm 2.5°$ in going from region to region. These data, along with the well-known twinning tendencies of ZnS, allow us to propose one possible model for the formation of canals depicted in Fig. 3. That figure is a schematic representation of a section of the closely-packed atomic layers in two (110)-plane ZnS crystal regions parallel to the crystal's optical axis. The layers of one of the regions are truncated by the edge of the other, forming an edge dislocation perpendicular to the plane of the drawing. For each such dislocation there is a rarefied region along which ion drift is significantly facilitated. Dislocations perpendicular to the optical axis in one of the regions are parallel over their entire boundary. In the other region the optical axis is rotated by an angle equal to the angle of birefringence. At the next boundary the picture is the same, except for the fact that the dislocations must be rotated by 60° around the optical axis, relative to the way its depicted in the figure. Therefore, no matter how the electric field is oriented perpendicular to the optical axis, there will still be a group of dislocation planes found in the crystal such that the direction of their edge dislocations is no more than 30° off the direction of the field. Impurities will drift along these inside the crystal.

In order to ascertain whether this description of doping is peculiar to our particular single crystals, grown from the vapor phase as described in [5], we conducted experiments using ZnS crystals grown by various methods: 1 — from a melt, 2 — from the vapor phase (the method we used), and 3 — by chemical transport (the modified Markov-Davydov method). In all three cases, after being knocked out in the same way, we still observed the interference bands on the crystal planes testifying to the regional structure of the crystals, and electrolysis occured at approximately the same rate when the electric current was fed along the $(111)_c$ plane. From this we can deduce that this description of the structural characteristics of ZnS follows from its tendency toward twinning and the existence of hexagonal interlayers (overlap flaws), and is present in all ZnS crystals grown at temperatures higher than the phase transition.

The existence of these sample-oriented channels, along which the drift of impurities takes place, confirms yet another experimental fact. When large quantities of NaCl are applied to the electrode, after electrolysis thin black bands are observed in the crystal. They are strongly oriented with the interference bands and go through the entire crystal from electrode to electrode, located as it happens in the several boundary areas between the interference bands. Accumulation of foreign particles in a dislocation plane will be visible only if their size (at least in two dimensions) is comparable or larger than the wavelength of light, or if their orientation plane is not close to along the observer's line of sight, i.e. just like the orientation of the dislocation planes in our case. Therefore we conclude that these bands seem to be the dislocation planes as defined by impurity decoration.

THE SLOW DOPING STAGE

As is apparent from the foregoing, the doping of impurity ions in a crystal under the influence of an electric field can be represented by two stages: rapid drift along dislocation channels and ordinary diffusion in the perpendicular direction with speed characteristic of the given temperature. Despite the speed of the first doping stage, this mechanism cannot provide a uniform distribution of impurities throughout the volume of the crystal if the diffusion coefficient of the impurity is not sufficiently high. Actually, the characteristic depth of regions of constant birefringence is, according to our data, in the tens of microns. As is apparent from [4], diffusion itself will take place under our experimental conditions to a depth of less than a micron. Thus the centers of the regions will remain unactivated. In addition, low-temperature diffusion will not, in general, produce the same kind of centers as will arise under activator doping during crystal growth at high temperatures. In fact, under such conditions an activator ion can easily find a suitable site (for example, near a compensating volume or charge in the intrinsic defect lattice or a co-activator ion) or even create such a site for itself (for example, promoting the formation of a nearby cation vacancy). This doubtless promotes an increased concentration of activators in the crystal, but without the accompanying experiment it's difficult to say whether concentration luminescent centers would also form in this process. Some associative centers can appear to be quenching centers. During activation of an already-prepared crystal at low temperatures the probability of formation of such associative centers is significantly less, since the crystal lattice is already formed and all the sites near the activator, introduced during crystal growth, are already occupied.

Therefore in our experiment, we consider the appearance of photoluminescence characteristic of a given activator to be an indicator that that activator in the required form has been introduced into the crystal. The original crystal wafers were either completely non-luminescent at room temperatures or displayed only weak photoluminescence in the form of diffuse bands encompassing the visible spectrum. Photoluminescence occurred immediately after electrolysis only when copper or silver were doped in. During activation by silver, the luminescence spectrum assumed a singular bell-shaped curve with a maximum at 2.7 eV (Fig. 4, curve 1). This and all following photoluminescence spectra were obtained at room temperature during pumping by light of 365 nm wavelength extracted from the radiation of a mercury lamp with filters.

During the electrolytic doping of copper, as many as three overlapping bands of luminescence with maximums at about 2.7, 2.3, and 1.8 eV were registered in the photoluminescence spectrum for a series of samples (Fig. 4, curve 2). The spectrum of a single band in the green was observed during electrolytic doping of copper after the initial crystal had a thin film of metallic copper vacuum-deposited on it (Fig. 4, curve 3).

Apparently, all these bands are linked to the presence of copper in the zinc sulfide lattice. Thus the red band with $h\nu_{max}=1.8$ eV can be associated with the red band of copper observed in powdered luminophors after electrolysis with copper at high temperatures [2,3], where these bands were ascribed to copper ions found in interstitial sites and were not associative with oxygen. The spectrum of these luminophors is reproduced in Fig. 5 (curve 4). This opinion is confirmed also by experiments on the electrolytic doping of copper into unactivated crystals obtained by severely restricting the flow of oxygen during growth. Wafers knocked out of these after electrolysis provided no strong luminescence in the red and indigo regions of the spectrum, and an almost total lack of luminescence in the green (Fig. 5, curve 2). The intensity of the red band increased sharply after supplemental firing at 900° C in a desiccating argon stream (Fig. 5, curve 3), at which time the spectrum remained similar to the spectrum of powdered ZnS—Cu. At the same time it is necessary to point out, that in the initial unactivated crystal broad, structureless bands in this region of the spectrum were also observed, which could be connected either with the intrinsic defects or with the uncontrolled impurities in the copper. Therefore there is still some probability that part of the red glow observed in the electrolytically activated samples is connected with the intrinsic defects or their association with copper ions.

146

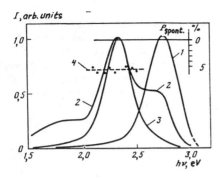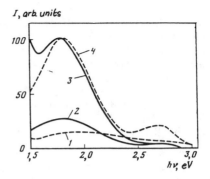

Fig. 4. Luminescence Spectrum of ZnS Crystals Activated by Various Elements with the Electrolytic Method. 1- Ag; 2-Cu from $CuSO_4$; 3-Cu from a metal film; 4- Level of spontaneous polarization at 300 K for the spectrum of curve 3. Optical pumping at 365 nm.

Fig. 5. Luminescence Spectra of Zinc Sulfide Samples Electrolytically Activated with Copper 1-Initial unactivated ZnS; 2-Same crystal after electrolysis with Cu at 600° C; 3-Same crystal after suplemental heat treatment in argon at 900° C; 4-Powdered luminophor ZnS-Cu [2,3]

The absence of a green copper glow in the samples obtained under oxygen-free conditions does not of itself indicate an absence of copper in the sample. The situation is, as was demonstrated in [6], that the usual green luminescence of copper in powdered zinc sulfide belongs to the associative centers Cu—O, which are maintained even after deoxidation of the crystalline phosphor in hydrogen sulfide at 1200° C, but are annihilated upon plastic deformation of the deoxidized powder. Thereupon the green luminescence band disappears at the same time as EPR of the Cu^{2+} ions appears, observable even at room temperatures. Apparently the green luminescent centers obtained during electrolysis have the same structure as those obtained in [6]. In any case, the fact that the observable part of the radiation in the green region was caused by copper is substantiated by the presence of characteristic negative spontaneous polarization of this radiation with $P_{spont}= $ —3 to —4%, as measured for the samples shown in Fig. 5, and $P_{spont}= $ —5 to —5.5% for samples activated with copper via metallic-copper electrolysis (Fig. 4, curves 3 and 4). These values are close to those found in [5] for crystals activated by copper during growth, wherein the green band was wholly due to radiation of the "copper" centers with spontaneous polarization of $P_{spont}= $ —8 to —10%.

This very topic, the appearence of activator luminescence of Ag and Cu in low-temperature firings was described by N. Ril' in [7]. He utilized a method whereby pure, well-crystallized, non-luminous zinc sulfide powder that had been previously fired at high temperatures was activated in a secondary firing at lower temperatures in contact with an activating salt. (Ril' called this the "contact" method.) In his work it was shown that during the firing, starting at 350° C for copper and 400° C for silver, one could detect luminescence characteristic of each of the activators, albeit weak, except for the red glow from copper which, as is now clear, was absent because their was no attempt at deoxidation of the initial powder. On the basis of this experiment, Ril' concluded that at such low temperatures the activator went to sites already "prepared" during the high-temperature firing

147

of the unactivated zinc sulfide. It may be that such sites are interstitial or cationic vacancies located near oxygen ions which have replaced sulfur. Since Ril' used powdered zinc sulfide in his experiments, it was enough for the activator to penetrate the crystalline particles only several microns deep to achieve complete activation. In our experiment, both during and after electrolysis , the same depth of impurity diffusion penetration also turned out to be sufficient for activating the entire volume of a crystal having significantly larger dimensions. This shows that our results agree with those obtained by Ril' via a different method.

The assumption of low mobility for the ions in the basis lattice at these temperatures (the negligible number of "prepared" sites being one consequence of this) was also corroborated by the fact that we could not by electrolysis lower by even a small amount the concentration of iron ions in the lattice sites which were present in large quantities in the melt-grown crystals. Thus the quantity of Fe^{3+}, which could be easily detected from its characteristic EPR signal, did not change at all after the crystal wafers had undergone three hours of electrolysis. From this one can conclude that, although in many cases electrolysis at 600° C can create "high-temperature" luminescence centers, it is unable to annihilate them.

For many other activators, these "prepared" crystal lattice sites in zinc sulfide apparently are not suitable, or the diffusion coefficient itself of these activators is too small. As such, according to the data in [7], the characteristic luminescence of Mn (the orange band) appears during activation by the indicated "contact" method only after subsequent firings at 800° C. This will be higher yet, apparently, for the rare-earth elements. Thus, as shown in our experiments, the rare-earth elements will not create their characteristic luminescent centers at the electrolysis temperatures we used (600° C), which manifests itself in the inability to excite the luminescence specific to them immediately after electrolysis. This does not signify, however, that rare-earth elements cannot be doped into a crystal by electrolysis.

To verify that they can all be doped into a crystal, we augmented the experiments with supplementary 1200° C firings of the crystals undergoing electrolysis. This second firing was always conducted without the field and with the same added treatment: after electrolysis the activating impurity was carefully cleaned off of the ZnS wafers and this part of the crystal was ground down 0.1 to 0.2 mm, after which the entire wafer was carefully flushed with alcohol. The clean wafer, with a layer of deoxidized ZnS now on all sides, was placed in a quartz boat , set in the long quartz tube stopped at one end, and the entire system was heated to 400° C in a desiccating Ar stream for 1.5 h, after which the temperature of the oven was raised to 1200° C and held there with injection of H_2S for 30 to 40 min. At the conclusion of firing the tube was removed from the oven and cooled rather rapidly in air. After this procedure, under ultraviolet excitation the wafers displayed luminescence with peaks and bands characteristic of the doped rare-earth elements. This radiation was uniformly distributed throughout the volume of the activated wafer. However, good results were obtained by the described procedure while working with melt-grown crystals having high concentrations of uncontrolled impurities. As we moved toward more "pure" crystals, those containing no foreign heavy metallic ion impurities, the photoluminescence after the second firing did arise, but the

148

brightness was considerably less and the resolution of the spectral lines worse than in the case of melt-grown crystals. This brings up the concept that several impurities in the "dirty" crystals (including

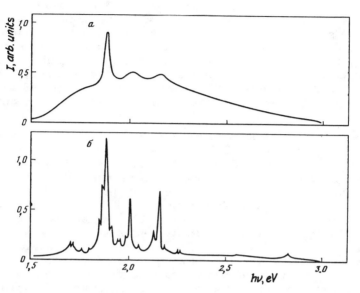

Fig. 6. Luminescence Spectrum of a ZnS Crystal Activated with Sm by the Electrolytic Method with Subsequent Thermal Treatment at 1200° C. a- without a coactivator; b- with Na as coactivator

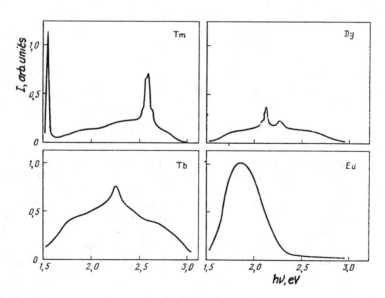

Fig. 7. Luminescence Spectra of ZnS Crystals Activated with Rare-Earth Elements by Electrolysis with Subsequent Thermal Treatment at 1200° C

oxygen) could act as coactivators. So we tried special doping of appropriate coactivators during electrolysis. For rare-earth elements, entering ZnS in the form of trivalent ions, we usually employed monovalent ions Na or Li as coactivators. Na (from NaCl) turned out to be better for preparation considerations in our experiment, since lithium salts are either difficultly soluble or very hygroscopic. All preparation procedures remained the same during the use of the coactivator. A small amount (one drop) of concentrated NaCl solution is added only to the solution for the activating impurities. Wafers prepared under these conditions, after electrolysis and subsequent high-temperature firings, showed significantly brighter luminescence than without the coactivator, and the line-spectra resolution became substantially better. This is demonstrated with the example of activation of ZnS wafers by samarium in Fig. 6. The same thing was done for activation of zinc sulfide by other rare-earth elements (Fig. 7).

The excellent activator line spectrum obtained in the process of such an operation (especially for Sm) indicates a lessening of nonuniform broadening, which in turn testifies to the fact that, in the presence of sodium, the rare-earth ions are removed from the dislocation channels, i.e. they go to an unfractured part of the crystal. Na^+ and RE^{3+} ions have almost the same radius which, as previously remarked, is larger than the radius of Zn^{2+} by approximately 17% and slightly higher than the limit set by the Goldschmidt rule. Then the positive role of sodium is apparently the result of charge compensation only of RE^{3+}. Sulfur-replacing oxygen ions can probably be compensating for the excess volume of both ions, since the radius of oxygen ions is significantly less than the ionic radius of sulfur. Na^+ and RE^{3+} ions substituting for zinc can have different charges relative to the ZnS lattice, since the former acts as an acceptor and the latter as a donor. Therefore they are attracted to each other. An oxygen ion is attracted to them by elastic forces, since the elastic tension in the ZnS crystal lattice is lowered as it approaches. So these three ions can form an associative center. However, during their doping into a crystal, at increasing temperature ($T = 1300°$ C) their dissociation is much more likely, and during electrolysis ionic mobility is low due to the significantly lower temperature. During the subsequent firing at 1200° C the Na^+ and RE^{3+} ions can easily appear to be lined up, because they enter the lattice through the very same dislocation, but in order to get to an oxygen ion they must penetrate a significant distance. Thus it is much more likely that for both methods, RE^{3+} will form both associative and single luminescent centers. However, this question goes beyond the scope of this article.

No matter what kind of centers the rare-earth elements form in zinc sulfide, our experiments show that electrolytic doping of ions of rare-earth elements in zinc sulfide crystals (with the intent of forming luminescent centers) should be conducted in two distinct operations: first it is necessary to introduce activator and coactivator ions into the crystal by elecrolysis at relatively low temperatures (600° C), and then conduct the high-temperature (1200° C) heat treating in a reducing atmosphere of H_2S (after switching off the field). This of itself opens up new possibilities for optimizing the activation of crystals and may be useful in the study of luminescent center structure.

TWO-ACTIVATOR CRYSTALS

It was shown in [5] by the electron paramagnetic resonance method that Eu^{2+} ions in single crystals of zinc sulfide without coactivators form centers with cubic and hexagonal (axial) symmetry of the crystal field. Eu^{2+} ions cannot associate in centers of cubic symmetry with a single volume-compensating impurity ion or with a single vacancy, since that would immediately lower the symmetry of the crystal field. In axial symmetry centers, generally speaking, this is not excluded. However, a case was given in [8] in which the lowering of the crystal field symmetry caused a field-induced interlayer or seam to appear with hexagonal close-packed layers in which these centers were found. This field has a definite direction perpendicular to the interlayer. It disappears upon plastic deformation, along with the hexagonal interlayer itself. The amplitude of the EPR lines belonging to these centers also simultaneously decreases, as does the degree of polarization of the luminescence. This shows that, if europium without coactivators is forming associative centers in zinc sulfide, then these centers are, in all probability, not luminescent centers.

All this reasoning, however, is based on circumstantial evidence, since a second impurity was not specially introduced into the crystal in [5,8]. Thus there is an interest in learning whether europium forms nonluminescent associative centers with impurities that are themselves activators in ZnS and can associate with europium. Copper was selected as such a second activator because it, like sodium, is an acceptor in ZnS. This activator was also chosen because ZnS—Eu,Cu crystals, which are activated during growth, are capable of electroluminescence when they have high copper content, with a luminescence spectrum close to that of ZnS—Eu, at a time when our preliminary experiments showed that EPR of Eu^{2+} ions was absent in it. It is then above all necessary to learn whether the luminescence characteristics of single crystal ZnS activated by two activators (Eu and Cu) by different methods will be unique: in the first case, activation by both of them simultaneously during growth and in the second case, activation by only one of them during growth (e.g., Eu) and by the other (e.g., Cu) in a separate operation at much lower temperatures. In powdered crystalline phosphors the second activator can be introduced by diffusion, but in single crystals this method will only introduce impurities to a negligible depth, significantly complicating the interpretation of the obtained results. Therefore in this instance we used the electrolytic method for doping of copper. Copper was introduced as described above. It turned out that, independent of the valence of the copper in the activating salt, a green band of $h\nu_{max} = 2.32$ eV appeared in the luminescence spectrum of the crystal after electrolysis (Fig. 8), and in the region of the Eu^{2+} bands the spectrum did not change appreciably. Meanwhile, a high degree of spontaneous polarization (20%) in the luminescence was maintained, and the EPR signal was reduced somewhat, by a factor of about 1.5 to 2. We should point out that, in fact, a somewhat slighter reduction (by a few percent) of the Eu^{2+} EPR signal was also produced in electrolysis of a controlled ZnS—Eu sample without activating copper impurities, i.e. the small decrease in EPR is connected with the heat treatment process in a neutral (Ar) atmosphere and is possibly a result of a partial restructuring of the zinc sulfide lattice — a change in the fraction of it which is hexagonal.

A completely different picture obtains when the crystal is activated by both activators during growth. In these crystals, even with concentration rations of the activators [Cu]:[Eu] = 1:10 (in the melt), the characteristic EPR signal for Eu^{2+} is strongly attenuated, and at [Cu]:[Eu] = 1:4 it disappears completely. However, as follows from the EPR observation of secondary impurities such as Fe^{3+}, which is present in significantly smaller quantities in these samples than Eu and Cu, the periodicity of the ZnS is not disrupted as a result of copper doping and thus the reduction in Eu^{2+} EPR is not a consequence of anisotropic line broadening.

At these concentrations of copper in the mixture, the polarization of the europium band luminescence decreases and then disappears completely. The maximum of the luminescence spectrum is shifted towards the red by approximately 0.1 eV compared to the spectrum of crystal ZnS—Eu activated by europium also during growth (Fig. 9 curves 1 and 2), and a new band appears in the region of $h\nu < 1.5$ eV. The spectrum of ZnS—Eu, Cu more

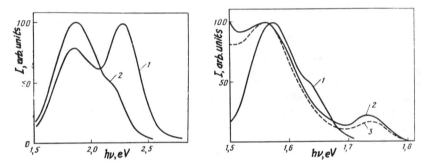

Fig. 8. Comparison of Luminescence Spectra of Crystals of ZnS—Eu and ZnS—Eu,Cu Activated by Different Methods 1- ZnS-Eu, activated during growth; 2- ZnS-Eu,Cu activated by electrolysis

Fig. 9. Comparison of Luminescence Spectra of ZnS Crystals Activated During the Growth Process 1- ZnS-Eu; 2- ZnS-Eu,Cu; 3- ZnS-Cu

resembles the spectrum of ZnS—Cu without Eu (curve 3) grown under conditions in which steps were taken to reduce the amount of oxygen in the air remaining in the apparatus during crystal growth. The same conditions were also necessary for obtaining the "red" luminescence of copper during electrolytic activation by copper of ZnS powder, as was already pointed out above (cf. Fig. 5, curve 4). It is obvious from the graph that all three samples have similar spectra with the exception of the very extreme part of the red, where one more band appears which we were unable to register on our spectroscopic apparatus.

From a comparison of the properties of the ZnS—Eu, Cu crystals obtained by these two means, one can conclude that, if there is copper present in the crystal contributing a green luminescence band (for example, after electrolytic doping without the subsequent 1200° C firing or during doping in sufficiently small concentrations at the time of crystal growth), then the distinctive features of Eu^{2+} centers (high luminescence polarization and EPR signal) are maintained. If the copper contributes a red band, then both the luminescence polarization and the Eu^{2+} ion EPR is absent. The 1200°

C firing after electrolytic doping of copper has the same effect as an increase in the copper concentration during crystal growth, but the effect is not so clearly expressed. Apparently, green luminescence centers are formed during doping of the zinc sulfide crystal lattice by copper which preferentially associates with oxygen. If they can't get sites near oxygen (because of too little oxygen or too much copper), then part of the copper ions remain in interstitial sites of the regular lattice to form red luminescent centers, and part associate with Eu^{2+}. The centers which this gives rise to do not exhibit EPR and, probably, do not luminesce (in any case, their luminescence spectrum differs strongly from that of the Eu^{2+} ions alone). This brings up the question, whether this effect will appear when there is several times more europium than copper. It is possible that the answer lies in the fact that the given concentration ratios of Cu and Eu are referenced to the melt, and not to the regular crystal lattice, where europium is less readily accepted than copper. However, neither a chemical nor a spectroscopic analysis will allow us to find out how the impurities are distributed between the regular lattice and the extended defects (for example, dislocations). So this question remains open.

REARRANGEMENT OF THE CRYSTAL STRUCTURE

As was noted during the experiment described above, firing of the wafers cut from ZnS crystals at both 1200° C and 600° C led to a change in their birefringence. This demonstrates that at such temperatures the proportions between the hexagonal and cubic phases of the crystal changes, i.e. a thorough rearrangement of its structure takes place. Firing at 1200° C always increases the birefringence, i.e. increases the depth or the quantity (or both) of the hexagonal interlayers. For example, one of the electrolysis-activated ZnS—Sm crystals had 2.7% hexagonal phase after electrolysis, and after 1200° C firing this amount grew to 32%, i.e. by a factor of 12. Usually we used crystals that started out with a percentage of hexagonal phase that was several times larger than that (15 to 20%), but even in those cases the 1200° C firings increased the percentage, although not as significantly. The increase in hexagonal phase content of ZnS—Eu could be followed from the increase in the EPR signal from hexagonal symmetry Eu^{2+} centers as compared to the sinal from centers of cubic symmetry.

The rapid growth of the hexagonal phase during such firings is evidently explained by the fact that the hexagonal phase itself is stable at these temperatures, and the phase equilibrium point is more than 100° C lower. Thus at 1200° the cubic phase of the crystal is already far from its equilibrium. Moreover, on account of the high temperatures, it is relatively easy to overcome the potential barriers preventing the propagation of edge dislocations which, inherently, also lead to a rearrangement of the crystal structure. The post-firing cooling in our experiment proceeded much faster than cooling after the growing of the crystal. Thus, in the time it takes for cooling from the equilibrium point between the cubic and hexagonal variants to the temperature at which the propagation of dislocations is practically completely "frozen", the crystal does not manage to rearrange completely from a cubic to a hexagonal lattice.

It is more difficult to explain why the 600° C firing causes a decrease in birefringence. As it is, the birefringence of ZnS—Eu crystals decreases by 25 to 30% of its initial value after a 600° C firing. This corresponds to a

153

decrease in the hexagonal phase fraction from 15 to 20% down to 10 to 15%, which is immediately expressed in the form of a "fluted" or "grooved" spectrum. Fig. 10 shows the "fluted" spectra of the same ZnS—Eu crystal that had passed through all the heat treatment stages. We note that that a

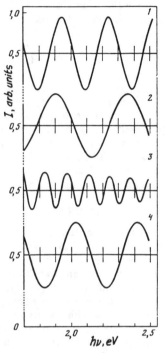

Fig. 10. Fluted Spectra of ZnS—Eu Crystals at Various Stages of Heat Treatment 1- Initial crystal; 2- After the first heat treatment (600° C); 3- After the second heat treatment (1200° C); 4- After the third heat treatment (600° C)

decrease of the hexagonal character as a result of the 600° C firing sets in even during heating of the sample for only a half-hour (the usual duration of electrolysis). At crystal cooling rates typical during growth processes in this temperature region (on the order of 100° C/h), this corresponds to a cooling of the crystal (over the time period of the repeated heatings) of only 50°. As such, this heat treatment is not equivalent to a simple delay in the cooling process.

Apparently the situation here is that the activation energy necessary for propagation of edge dislocations depends on whether or not they have to "jump over" impurity atoms. The propagation of just these dislocations causes a reorganization of the hexagonal lattice into a cubic one at 600° C (and propagation to the opposite side causes a reorganization in the reverse direction at 1200° C). If the impurity atoms are hindering the propagation of dislocations, then they will delay them all the more in their vicinity the lower the temperature. If there is a large enough difference in the activation energies for propagation of dislocations in the regular lattice and for their "jumping over" impurities, then it may turn out at some temperature that they can still propagate through the regular lattice, yet not be able to "jump over" impurity atoms. Apparently this is what happens at 500 to 700° C during cooling of the newly-grown crystal. At these temperatures, complete lattice reorganization is still not possible, but in the lattice elastic stresses are accumulated in dislocation pile-up sites around impurities. This lowers the potential barriers which dislocations must overcome if they are to continue propagating. The crystal, having been heated once again to 600° C, is not in the same state it was at that same temperature during the post-growth cooling, because now elastic energy has accumulated in it. Apparently the thermal energy at 600° C is already enough for the elastic-stress-induced dislocations to overcome the potential barriers standing in their way. They begin to propagate, causing a local reorganization of the lattice, but simultaneously lowering the elastic stress. In the end, these stresses are relieved and again the propagation of the dislocations is stopped. Thus the second firing at 600° C (cf. Fig. 10, curve 4), while lowering the hexagonal phase fraction in the crystal, does not bring it to zero.

The role played in this by the impurities themselves is apparent from the following experiment. In a ZnS—Sm crystal, the percent hexagonal phase increased after a high-temperature firing from 2.7 to 32%. A second identical crystal which had not been doped with samarium was simultaneously subjected to the same firing. In it, the hexagonal phase content did increase, but only by 10%. From the previously described point of view, it can be explained that, thanks to the lower impurity concentration, the dislocations can propagate in it more freely and thus the transformation from hexagonal to cubic phases during cooling will be more complete.

A detailed investigation of these processes was not conducted in our inquiry. However, from the foregoing observations it would follow incidentally that, apparently, low-temperature annealing is more "invasive" than we realize. It can change the proportion between the hexagonal and cubic phases in the zinc sulfide lattice, and not only lower internal stresses, but under some circumstances increase them as well.

CONCLUSIONS

Thus, although the electrolytic method has turned out to be more complex than it appeared at the start of our research, it has proved to be fruitful, especially in two-activator doping. Comparison of the properties of crystals which were doped with the same activator either during growth or during lower-temperature electrolysis is also promising for investigation of luminescent centers. We did not suceed in showing the luminescence bands characteristic of the association of activator ions with the intrinsic defect lattice. However, if such associative centers do indeed exist, we can discover them by means of just such a comparison (in part, comparing the fine structure of the the spectra of trivalent rare-earth ions), since the relative concentration of associative centers doubtlessly depends on the method of activator doping.

There is also interest in extending the electrolytic method of activation to other crystals. It should be borne in mind that, even if the crystal contains very few mis-packed layers and consequently few channels formed by dislocation faces, one still can find favorable directions for the electrolytic introduction of impurities. This may be the case, for example, in the well-known channels in hexagonal lattices of the CdS type, which are parallel to the C axis.

In conclusion, we would like to express our gratitude to A. G. Glyadelkina, A. V. Lavrov, and L. M. Tsyganova for growing the majority of the crystals especially for this research and carrying out the high-temperature treatments, to N. A. Gorbacheva for growing unactivated crystals by the modified Griyo method, and also to I. K. Vasilenko, Yu. V. Korostelin and P. V. Shapkin for loaning us unactivated crystals grown by transport in the modified Markov—Davydov method.

BIBLIOGRAPHY

1. Andreyev, I. S., Zyrina, L. V. and Arzyman'yan, G. V., "Electrolysis as a Method of Activating Electroluminophors", IZVESTIYA AN UzSSR, No 4, 1961, pp 83-86.

2. Voznesenskaya, T. I. and Fok, M. V., "Orange Phosphor ZnS—Cu Obtained by the Electrolytic Method", OPTIKA I SPEKTROSKOPIYA, Vol 18 No 2, 1963, pp 249-252.

3. Voznesenskaya, T. I. and Fok, M. V., "The Nature of Red Luminescence in ZnS—Cu Phosphors", OPTIKA I SPEKTROSKOPIYA, Vol 18 No 4, 1965, pp 656-660.

4. Bochkov, Yu. V. and Georgoviani, A. N., "Luminescence and Photoelectric Characteristics of Zinc Sulfide Diodes", TRUDY FIAN, Vol 138, 1983, pp 46-72.

5. Arkhangel'skiy, G. Ye., Bukke, Ye. Ye. and Voznesenskaya, T. I., "Visualization of the Fracture Structure in Crystals of the ZnS Type by the Decoration Method", TRUDY FIAN, Vol 164, 1985, p 103-113.

6. Arkhangel'skiy, G. Ye., Grigor'yev, N. N., Lavrov, A. N. et al, "Transformation of Luminescent Centers in ZnS—Cu, O During Inelastic Deformation", TRUDY FIAN, Vol 164, 1985, pp 103-113.

7. Ril', N., "Lyuminestsentsiya" [Luminescence], OGIZ, Moscow/Leningrad, 1946, pp 105-114.

8. Arkhangel'skiy, G. Ye., Grigor'yev, N. N., Fok, M. V., et al, "The Influence of Plastic Deformation on Luminescence and Electron Paramagnetic Resonance in ZnS—Eu Crystals", TRUDY FIAN, Vol 164, 1985, pp 42-103.

SUBJECT INDEX

157